SPIRITUAL BODY OR PHYSICAL SPIRIT

Bio Dark-Matter Chemistry & Your Invisible Doppelganger

Philip P. Benjamin, PhD MSc MA

Spiritual Body or Physical Spirit

For information about special discounts for bulk purchases, please contact Sunbury Press, Inc. Wholesale Dept. at (855) 338-8359 or orders@sunburypress.com.

To request one of our authors for speaking engagements or book signings, please contact Sunbury Press, Inc. Publicity Dept. at publicity@sunburypress.com.

FIRST SUNBURY PRESS EDITION
Printed in the United States of America
February 2013

Trade paperback ISBN: 978-1-62006-182-4
Mobipocket format (Kindle) ISBN: 978-1-62006-183-1
ePub format (Nook) ISBN: 978-1-62006-184-8

Published by:
Sunbury Press
Mechanicsburg, PA
www.sunburypress.com

Mechanicsburg, Pennsylvania USA

IN MEMORIAM

In loving memory of Ivy Thomas Benjamin, M.D., over four short decades of life together, whose invisible soma awaits to be energized into the full glory of luminosity, and without whom this work would not have come about.

http://www.sciencedirect.com/science/article/pii/S0046817781801347

Dedicated to the Evidentialist (Doubting) Apostle Thomas fulfilling the post Resurrection Great Commssion by his mission to the Cochin Jews and others in A.D 52, martyred at Mylapore (Madras) in A. D 72 and establishing seven churches, the traditional roots of the Palamutil and all the other branches of the Sankaramangalam (Sankarapuri) family tree in Malankara (Kerala, India).

http://www.newworldencyclopedia.org/entry/Saint_Thomas_Christians
http://kottayamcheriapally.com/466/511/about-church/history/history-of-orthodox-church

Contents

Key words: Electromagnetic (EM), non-EM, biophotons, monopoles, anomalous events, bio dark-matter (BDM), bio dark-matter chemistry, light and dark chemical bonds, spin, dark-matter body (DMB)

Foreword

From the perspective of explaining both local and non-local human consciousness, the essential ingredient missing is the coupling mechanism between an axion condensate and the human brain. Neither the Iskatov's theory nor Jerome's model has any provision for that. Furthermore, the axion particle has yet to be detected, though there is strong consensus that some sort of pervasive, global quantum coherent medium must exist, and the most likely candidate is the theoretically derived axion condensate. It would then appear that the next essential step is to determine coupling mechanisms between the condensate and the physical brain. Paavo Pylkkanen reviewing Bohm pilot wave theory relates to the coupling problem "of mental causation, or the problem of how could mental states (as non-physical states) possibly influence the course of physical processes without, for example, violating the energy conservation laws. Bohm and Hiley's ontological interpretation of quantum theory suggests that an entirely new kind of energy operates in situations where quantum theory is required, an energy best understood as 'active information' which exists as 'pilot waves' (as wave functions) enabling response of fundamental particles such as electrons to pilot wave information. There is no wave/particle duality. They both exist at the same time and have definite locations in space and time. http://members.aol.com/Mszlazak/BOHM.html, However, Bohm's opinion that BECs could be conscious cannot be supported, rather BECs may couple the mind to the brain. It is more likely that Bohm's pilot waves exist in the axion condensate of Dark Matter. But the question remains as to how the particles get propulsion to change course. The reader is reminded that essentially the same thing must happen in the frictionless flow of electrons as Cooper-pairs through superconductors. Somehow the pairs move around the obstacles of atoms. So either there are unknown forces in the BEC of the superconductor to guide dumb particles; or the fundamental particles are smart and can sense the environment and change direction without loss of energy. Philip offers a means to do this coupling via bio dark-matterchemistry and suggests that the *resonance* between the two may be the basis for self-awareness, since resonance is rudimentary *recognition.* He stipulates that the "how and where" of that transduction is not a chemical problem and leaves that to physics and neurosciences, citing the work of Niemark et al (2003) at Harvard as an analogy, who present evidence for "vibrissa resonance as a transduction mechanism for tactile encoding" in rats. His argument is compelling, that from the point of view of physical sciences, only the physical can interact with the physical, and that since self exists and functions as a realityin the physical body, it must necessarily be physical. Thus, if "self" is invisibly functioning, then it must belong to a category of invisible matter. As the visible physical body is electromagnetic (EM), this invisible physical "self" must be non-EM. Then the EM body is coupled to and permeated by a non-EM body. Though for a while he had toyed with the idea of back-action quanta, he concluded that the only viable candidates for the production of that category of a physical body will be bio dark-matter and its chemistry, co-creating the dark-body from the moment of conception via processes subject to genetics. The non-EM bodyfor all practical purposes has all the mystifying qualities of the supposed mystic consciousness, but unlike the latter it is physical and subject to scientific analysis. Some years ago, Chris King first pointed out to me that BECs are too simple to support consciousness. That seems to be a mute point, because it is not the "simple" BECs which are the actors here, but the very complex configurations of dark-matter particles coupled to the equally complex combinations of light-matter particles.

The intriguing taxonomic gradation of sentience compelled him to look for a measurable physical factor which is common to all species but vastly different in values across the taxa. That search led him to the experimental data on biophoton (ultra-weak photon) emission which is intrinsic to all live matter, but vastly different from plants to humans by an order of magnitude. He contends logically that dark-matter body is differently constituted in different species. In humans it is verisimilar to the "body" that possesses it and is permeated by and integrally coupled to it.

-Richard Requist, PhD (Harvard)
Retired Physicist

Philip uses qualities of dark-matter to unravel the invisible but experientially present "*self*" in a rational way enabling future quantification in practice and therefore theory. As he elegantly puts it "where matter exists chemistry exists". Dark-matter gives dark-matter chemistry and a dark-matter body." His methodical and evidential approach including the analysis of emission of biophotons "*the light of life*" delimits our cognizable limits--the limits rationalism's hubris ignores to the point of ungrounded myth and myst about mind. Here is reason that does not exploit the mysterious derivation of matter, i.e., life as such. Philip penetrates matter (though constantly aware of our cognizable-predicament) and participates in a healthy, informed cautiousness toward using philosophical reasoning (but only vs. metaphysical rationalism)—an existential attitude made clear in the positive theistic existentialistic works of Karl Jaspers. There's the leaning toward a reasoning that encompasses and shatters rationalism. While the source of life is implied, he respectfully avoids eye contact and meaning by association with the divine. It's an extraordinary unifying quiet approach that cannot be reduced to a dimension that excludes occidental historical wisdom. The work presents a friendly unifying natural view in so far as it preserves an open-endedness toward experience.

Glenn Wood, BSL. MA
Researche, Karl Jaspers Applied

[Philip's] work is objective and well written and [the] logical stipulation on the role of axions and/or equivalent concentrates should appeal to any neutral reader.Bio dark-matter chemistry is needed for the existence of bio dark-matter DNA/RNA intermediate receptor influencing the ongoing genetic transcription process that results in the production of the relevant and adaptive m-RNA à enzyme in the decision-making process (my Vol. IV of Neurophilosophy of Consciousness). There is no model theory in existence, whether Iskatov's or Jerome's, that bridges the external source and the physical brain. Here I call for a new paradigm shift, a head on plunge into dark matter chemistry, in order to explain how the so called 'invisible' mind is causally efficient in interacting with the visible body/brain to produce measurable mesoscopic activities including biophoton emissions. I endorse the value of [the author's] presentation.

-Angell O. de la Sierra, PhD, JD. Esq.

Prolegomenon

The Gist of the Argument

A casual encounter with a Boeing engineer at a hair salon led to a *tête-à-tête* on the stuff that makes natural laws possible. If visible ordinary matter (*light-matter*) gives ordinary chemistry and a visible body, then invisible extraordinary matter (*dark-matter*) will give extraordinary chemistry and an invisible body . Ordinary matter is electromagnetic (**EM**). It has single point ***electric charges***. Then extraordinary matter is **non-EM.** It will have ***monopoles*** which are single point magnetic poles. Evidence is mounting that the biosphere has both **EM** and **non-EM** matter, especially from ultra-weak photon emission which is intrinsic to living matter only. If there were no laws of physics and chemistry, electrons (**e**), protons (**p**) and neutrons (**n**) will be still whizzing around with no possibility of the existence of a biosphere as we know it. Suppose these **EM** particles, **e, p, n** have parallel to them **non-EM** *dark* particles, say, **E**, **P**, **N**. Then they also will have laws of physics and chemistry. The **non-EM** aspect makes it invisible and hence the name **dark matter (DM)**. For the biosphere it may be called **bio dark-matter (BDM)**. It may or may not be the same as **cosmic dark matter**. Bio dark-matter can stand on its own with or without any relevance to the cosmological dark matter. Evidence for the existence of dark matter is abounding in theoretical physics and astrophysical calculations. Theoretically determined axions have been suggested as dark-matter. Matter without chemistry is not of much use in the biosphere. That suggests the need to explore the feasibility of dark-matter chemistry, possibly involving monopoles and spin similar to charge and spin in ordinary chemistry. It could be neutrinos or the theoretical axions. The evidence for monopoles is also increasing. *Bio dark-matter chemistry* involving monopoles is in question. The nature of the invisible bio dark-matter restricts experiments to *ab initio* computational chemistry only. When laboratory experimentation is impractical, computational resources present a practical alternative for the combinatorial possibilities of synthesizing novel molecules from fundamental elements, which has now become an ardently sought after holy grail of computer science. *Ab initio* computations will be the only method available to explore *bio dark-matter chemistry.*

There will be two sets of physical bodies in the biosphere, one is observable and made of **EM** particles. The other is unobservable and made of **non-EM** particles. A visible *light* body will thus be joined to an invisible *dark* body via *spin-spin coupling.* Though **EM** and **non-EM** matter have very little direct interaction, chemical bonds belonging to each can weakly interact causing orbital energy changes that lead to low intensity photon emission. Such photons will be very specific to living matter only. In fact they do exist and are called biophotons (first observed by Alexander Gurvich in Russia, 1923). The emission rates differ vastly almost by an order of magnitude across the taxa, plant cells emitting about 10 times more than human cells per sq. cm/second. This indicates different degrees of completeness of dark matter bodies from plants to animals. Plant cells may have only one kind of dark matter particles parallel to electrons only, while human cells may have three different kinds of dark mater particles corresponding to **e, p, n**. A **bio dark-matter** body, an invisible doppelganger, will be relatively at a negative energy state **($-E=mC^2$**, m= body mass) when decoupled from the visible body of mass ***m*** at death. In such a dormant state there is no possibility of a roving ghost, unless energized from powerful external sources. Oh, that rings a bell- of resurrection, said the Boeing man. Then the tonsorial summons terminated that chitchat.

Spirit & Soul

During freshman days when electrons (e), protons (p) and neutrons (n) came across as the basic constituents of all life forms, a stupid notion had stuck in my mind (or head?) that the commonly believed invisible *spirits* may then be made of corresponding invisible matter. In later years those vain impressions were translated into emoton, psychon and neumaton as the particles corresponding to e, p, n. A philosophy journal rejected that paper on the grounds that it is more science than philosophy. Then I came across a post in Quantum Mind positing that mind is a phenomenon of *quantum back-action* of presumably electrons. I added to it the *quantum back-actions* of protons and neutrons to form a quantum *back-action human body (Jul 14, 1998. RECD 97. http://listserv.arizona.edu/cgi-bin/wa item# 000035 http://listserv.arizona.edu/cgi-bin/wa?A2=ind9807&L=QUANTUM-MIND&P=R1552&I=-3).* Responses to that stimulated a search for more evidential objects which eventually led to the chemistry of *bio dark-matter,* with biophotons from *light* and *dark* chemical (bond) interactions as indirect evidence, since *dark-matter* has little or no direct interaction with *light- matter.*

If *spirit* and *soul* are mystic mysteries, then they are outside the scope of physical sciences. All physical events must have sufficient causes that lie within the physical world. For a mental event to be anything other than a physical event, a violation of physical law is required to make any causal contribution of its own in the physical realm. **Soul and spirit** in all ancient languages have the roots as breath and wind. However, mysticism does not allow that **soul/spirit** is made of wind! So something non-physical had to be invented. That still cannot explain why or how the non-physical can be sentient. Introducing the term *consciousness* from esoteric religions adds no benefits. It is no less mysterious than mind or soul or spirit. Nobody knows what this New Age *consciousness* is. Conscious and unconscious states are brain states, well within the scope of and very well explored by neuroscience and anesthesiology, but the mystic *consciousness* is not. Religious or metaphysical meaning has been obviously added to the "brain states of consciousness" and *consciousness* is apparently the mind or spirit or soul wrapped in Eastern mysticism. Its origin can be traced to the Tao physicists and psychologists involved in the Copenhagen interpretation (rather *Copenpagan imagination*) of quantum mechanics. Interpretations are not necessarily science; often they belong to the realm of metaphysics. Such quantum vanities have produced many a psychobabble.

Materialistic reductionism is rooted in ordinary matter. It requires that both living and dead matter be viewed as mere concatenations of electrons, protons and neutrons. That view has inherent limitations: **1. Taxonomic conundrum**. Plants, animals and humans as well as the dead matter around them are all essentially made of the same physical particles. If one is sentient others must also be sentient, so also for sapience. However such is not the case empirically, experientially and experimentally. The issue of panpsychism cannot be dismissed with, "so what"? Or rocks may be sentient, who are you to deny that? Panpsychism is still only on a scientific wish list. **2. Anomalous phenomena**. Out of body and near death experiences (**OBEs, NDEs**) have been reported from time immemorial. Here two distinct kinds of physical matter seem to be involved. Since ordinary matter is electromagnetic (**EM**), the other matter could very well be **non-EM,** physical nonetheless. **3. Biophotons** and their taxon dependent emission rates. Ordinary materialism has no clue for either the origin or the vast difference in emission rates of biophotons, by an order of magnitude, across the taxa.

Mystic *consciousness* supposedly collapses a quantum mechanically derived mathematical wave-function, which then creates *consciousness* which again collapses the wave-function etc. Thus the circular reasoning goes on ad infinitum. All this wishfully occurs in some capillary tubes of optimum dimensions (microtubules) in the brain. There seems to be secular mystics and religious mystics committed to these notions. Those who proffer Biblical validation of mystical states completely ignore the physical aspects of resurrection which the Bible affirms. An energized **non-EM** physical body can possess the characteristic properties of the resurrected body reported by eyewitnesses in the Bible. It can easily permeate or penetrate any **EM** barrier. A **non-EM** body is enduring. It may be highly radiant or ghostly visible, depending on its energy state. When decoupled upon death it has to be externally energized from a humongous source, if it is to be functional.

Benjamin Baird and Jonathan Schooler, psychologists at the University of California, Santa Barbara report in Psychological Science (in the press, May 2012): "We've traditionally found that rapid-eye-movement sleep grants creative insight. That allowing the mind to wander does the same is absolutely fascinating". John Kounios, a psychologist at Drexel University in Philadelphia, Pennsylvania, agrees: "This finding really plugs a hole in the literature". They are reporting this as ***interesting finding*** without daring to draw any conclusion other than that mind-wandering seems totally counterproductive from an *evolutionary* perspective, and has been viewed as dysfunctional because it compromises people's performance in physical activities". So they assume that if it is not genetically determined, "selection" advantages will explain it. What exactly do they mean by "mind wandering"? Is it minimum neuronal firings? Since they identify brain with mind, what rational argument can be presented to substantiate the superficial claim that a "dormant" or "inactive" or "wandering mind" can be very creative? Dormant brain is just that- dormant. It seems more likely that while the visible physical brain is resting or dormant, there is a more enduring invisible physical brain which continues to be active. The 'constantly dying' visible brain is necessarily less efficient. When this is temporarily out of commission and out of the way, the more efficient invisible physical brain becomes immensely active and creative, producing the results which the visible physical brain could never have dreamt of. If the visible physical brain is made of electromagnetic (**EM**) particles, then the invisible physical brain is made of **non-EM** particles for which axion-like particles or neutrinos (three flavors) and/or their monopoles are reasonably sound candidates. As far as physical sciences are concerned, *non-physical* is an oxymoron. A triune system of a neutrino body permeating an axion body which is permeating an ordinary fermion body (made of electrons, protons and neutrons) is more in line with what the ancients called the triunity of **spirit-soul-body** or the Vedic **paramatman, atman, sarir**. It resembles Plato's tripartite soul (**rational, spirited**, and **appetitive**), or the three gunnas of the

Upanishads [Sat (Pure), Raja (Passionate), Tama (Base)]; or the Buddhist normal level (sariri manussa), the supernormal (uttari manussa) and the sublime (ariya manussa). Freud formulated it as **Id, Ego, Super-ego**. Bible refers to the unity of **spirit, soul and body**.

Mysticism is a subjective experience, trying to separate the "**I**" or **self** from the body and join the **self** subjectively with a Cosmic **Self**. That is the essence of Eastern and Gnostic mysticism. Some believe that the dimensions of the microtubules in the brain are optimum for these quantum events to occur. Others point to enormous number of spatio-temporal dimensions (which are imaginary configurations of algebraic co-ordinates; these extra dimensions in Planck units being compactified/cubic centimeter). Physics wraps *consciousness* into the abstract quantum mechanical wave function or string-theory dimensions, a top-down approach to solve the mind-body problem. Empirically "**I**" or the "**self**" exists within the body. It is not something in my brain, any more than my fingers over a key board. The totality of my visible and constantly dying body cannot account for the continuance of this "**I**". Experientially, this "**I**" must be an invisible and durable physical body, a mirror image of the visible physical body, the two being identical twins co-created at the moment of conception. The **resonanc**e between the two may be the basis for **self-awareness** (Neimark *et al, 2003*). Resonance is rudimentary **recognition**. How and where that resonance is transduced into awareness is not a chemical problem. It belongs to physics and neurosciences. Since **self** exists and functions as a **reality** in the physical body, it must necessarily be physical. From the point of view of physical sciences, only the physical can interact with the physical. Thus if "**self**" is invisibly functioning, then it must belong to a category of invisible matter. As the visible **physical body** is electromagnetic **(EM)**, this invisible physical "**self**" must be **non-EM**. Then the **EM** body is coupled to and permeated by a **non-EM body**. The only candidates for the production of that category of a physical body will be bio dark-matter and its chemistry, co-creating the dark-body from the moment of conception via processes subject to genetics. The **non-EM body** for all practical purposes has all the mystifying qualities of the supposed mystic **consciousness**, but unlike the latter it is physical and subject to scientific analysis. The mysterious **biophotons** (ultra-weak photons) intrinsic to live matter may provide indirect evidences for the existence of bio dark-matter (Benjamin, 2003, 207 June & September, 2010 June and July, 2012)

It is commonly assumed that **mind** is spiritual meaning immaterial and **body** is physical meaning material. Spirit is shadowy and ethereal. That is mostly an Eastern concept. In the Western Biblical tradition, the resurrected body is visible tangible and physical and easily passable through closed doors. That is a characteristic of **non-EM** matter. The Pauline world view is not metaphysical, nor mystical, but real and observable, including the "third heaven". In Sanskrit, Greek, Hebrew and Latin, the root for soul and spirit is breath or wind. When the last breath was gone, the body was dead. So they determined that there is something "breath-like", invisible that keeps the body alive. Had the ancients been aware of "dark matter", most probably soul/spirit would have been called "dark" bodies! In Aristotelian physics, 'form' and matter are perpetually paired and the archetypical examples of forms are 'substances'. Matter and form are the "causes" of whatever exists. Incorporation of dark matter will replace the idea of a "breath-like" entity with an invisible dark matter body which will be more durable being nonelectric and non-entropic. Had they known "dark matter", soul and spirit would be "dark matter bodies". Metzinger's (2003), objection to **self** as a *non-physical soul-substance* is irrelevant since material objects can *physically* exist either *visibly* as made of *light-matter* (electric) or *invisibly* as of *dark matter* (nonelectromagnetic) particles. Besides, empirical investigations into the pathology of **self**, its development, structure, function etc, operationally needs the *reality* of **self**.

Over a thousand years before Paul, way ahead of Vedic mysticism the patriarch Job asked: "When a man dies, shall he live again?" He affirms: "Yet in my **flesh** I shall see Him" (Job 14: 13-15; 19:26-27). Thy dead men shall live, together with my dead body shall they arise (Is. 26:19). "Multitudes who sleep in the dust of the earth will awake: some to everlasting life, others to shame and everlasting contempt. Those who are wise will shine like the brightness of the heavens" (Dan. 12: 1-3). Paul uses the term *anastasis* for resurrection which refers to bodily rising (Rom. 1:4, 6:5; 1 Cor. 15; Phil. 3:10). Any assumption that **soul** and **spirit** are non-physical does not agree with the Biblical view of a physical resurrection- not metaphysical or mystical.

Introduction

The present motivation to incorporate dark matter to the study of consciousness comes from the recognition of the necessity of an ontic rather than an epistemic explanation of the link between the *subjective self* and the *objective body* via dark-matter chemistry. The bio dark-matter posited here is not necessarily the astrophysical dark-matter or some nebulous, ethereal, metaphysical entity, but real, physical, discrete charge-less axion-like particles. They are as diverse, dynamic and creative as electrons, protons and neutrons of the visible ordinary matter. The ontic randomness of quantum events is inadequate to provide space for mental causation. Most approaches to mind body problem based on ordinary matter exceeds the frontiers of present quantum theories and are highly speculative. Equating the activity of neuronal assemblies as the neural correlates of mental representations, offers no conceptual distinction between brain states and mental states, or any sound formal framework or concrete empirical scenarios for mind-brain correlations. If the biosphere and the rest of the universe are made of electrons, protons and neutrons only, then panpsychism necessarily exists, for which there is no evidence. Bio dark- matter chemistry will offer a fundamentally holistic, psychophysically neutral level of reality from with correlated mental and material domains. The dark-matter body with negligible mass permeates the entire visible body.

As early as the 6th century B.C. Thales and Anaximander in Miletus, introduced a rationalistic approach, instead of mythical accounts, to explain reality. The atomism of Leucippus and Democritus in 5th century B.C. and Aristotle's metaphysics also imply material substance as the fundamental constituent of reality. Concepts of substance as reality also emerged as a result of the Newtonian view of a deterministic cosmology with clock-like precision. Basic to the understanding of any material process is a quantum theoretical view of ordinary matter. None of these views could envision a non-electromagnetic dark matter component. Current science allows for the inclusion of the extraordinary dark matter also in any materialistic comprehension of reality. Bio dark chemistry presents a 'ball-and-stick' chemical scheme that will better explain life processes including the elusive **self**, than ordinary matter with the presumed unitary time evolution of quantum processes entailing *matter-wave duality*. An invisible *human machine* of non-luminous matter, functioning parallel to the visible and in resonance with it, will provide sufficient explanatory resources to account for all aspects of the invisible **self** and perhaps the mind-body problem. Thoughts become *encoded parallel changes* in the invisible neuron configurations. Awareness results from the interaction between the dark and the visible parts. **Self-awareness** arises from a dynamic process involving mutually *recognizable* entangled structures in *resonance* between the visible body and the physical invisible body parallel to it. Resonance is recognition. A substance ontology is capable of providing adequate physical underpinnings for a comprehensive theory of mind and mental processes, if we conjugate the luminous (**EM**) body with a parallel non-luminous (non-EM) counterpart. The luminous matter confirms the processes of becoming, and the more permanent non-electric dark matter corroborates the being. We can then examine the ontological status of ordinary molecules coupled to dark matter molecules and lay the groundwork for a conceptual framework which draws on a natural, mutually reinforcing quantum relationship between ordinary and dark matter chemistries. That leads to an intriguing concept of an invisible molecular physicality coupled to the visible, akin to mind coupled to the body. This is a powerful metaphor to model the science and philosophy about mental states, permitting to identify them with creative changes involving resonant molecular states of both visible and invisible matter.

Reality is not a mere mathematical function or a metaphysical process. The fundamental objects that populate the real world and their complexities endure through time. These permanent bits and pieces of our subjectivity are not mere patterns of chronological events stored in the ***electromagnetic (EM)*** realm of neurons and their correlates. The stable composite concatenation of wave-like quantum parts and the emerging sequential patterns and processes of ordinary matter, has only an appearance of permanence, offering an idealist speculation that the reality of mind is a mere derivative of this apparent constancy of particles. However over 80% of the matter in the universe is missing as *dark-matter*. It is *non-electromagnetic* **(*non-EM*).** It may even have *dark photons* also, perhaps as a force carrying particle. It is unwarranted to assume that dark-matter has no role in the biosphere. That role has to come through bonded states, as in the case of chemical bonds in ordinary matter.

Where matter exists, chemistry exists. Dark matter can have its own dark chemistry leading to *light* (EM) components of living organisms piggybacked by dark (non-EM) counterparts. Bio dark-matter can stand on its own with or without any relevance to the cosmological dark matter. It could be neutrinos or the theoretical axions. Acharya et al (2012) make broad predictions for a spectrum of axions as dark matter with a huge mass range from m(a)~ 10^{-33} eV to m(a)~ 1 eV. A bio dark-matter chemistry will involve monoploes. The evidence for monopoles is increasing. A dynamic unity of *light* and *dark* structures in resonance will offer a logical physical framework for many invisible but physical realities beyond the ordinary atomic world. The wave-like properties of dark macro-bodies are significant, since they have relatively small masses. A dark-matter body corresponding to a 125 Kg human body will be of negligible weight. Genetic information encoded on both **EM** and **non-EM** bio-systems will naturally select and govern all the bio-processes.

Personal annihilation is a frightening prospect to most people; almost any route other than death as final is preferable. The only atheistic hypothesis of after-life is naturalistic reincarnation. Buddhism, which is a non-theistic religion, propounds that tenet. But what is it that reincarnates? The body with its brain and all ordinary molecular structures disintegrate after death. Naturalistic reincarnation is an oxymoron in the entropic world of electromagnetic structures. However, one has to face the inevitability of death, the dark valley of its shadow being present from the very moment of conception. Oncogenes (cancer genes) exist naturally in all animals including humans. Development of cancer is normally a matter of time. Also, the telomeres at the ends of our DNA strands unravel at fairly regular intervals of cellular reproduction. Thus life in eukaryotic, multicellular organisms is literally programmed to die. The program can be accelerated or slowed down to a certain degree by mechanisms that are sometimes known and sometimes unknown. The Bible says: "You shall die". The "*sentence of death*" is imprinted on the genes. "Life" here exists in a physical body not in hyper-space. If there is life hereafter, that will also exist in a body, as an enduring non-entropic dark-matter body. The bio dark-matter body is left dormant on death at a relatively negative energy state by the decoupling of the visible body ($-E = mC^2$). This dormancy is instantly overcome when energized from a vast external source, resulting in a radiant tangible body.

If ordinary materialism is all there is to life, and the conscious **self** is established by the structures and processes within the brain, then **after-life** will be what **prior-birth** was, as if one never had a life. Then the **self** ceases to exist when these structures are destroyed. If the human mind is purely a byproduct of the ordinary quantum particles, then panpsychism is inevitable for which there is no empirical or experimental or experiential evidence. Prof. Hauser's (2008) 'humaniqueness' is too obvious, and the claim of the late Prof. Gould of Harvard that there is "no difference between twigs & humans", as all are made of the same stuff is devoid of evidence. From the perspective of ordinary matter, biology, neurology and brain functions all point to the likelihood that the marvelous brain has to have everything needed to establish a conscious **self**. In fact, an organism that survives by its mobility must be *necessarily* conscious in order to perceive, make decisions, use survival skills and find mates. Plants are not motile and thus do not have these needs. Awareness is much more sophisticated in mammals than in other species and most advanced with **self**-awareness in humans.

For too long, Science had to sideline *mind* because even a pragmatic definition of mind is unrealistic from the perspective of its *ordinary* materialist premises. Serious research into human consciousness has now become conventional, with many well-known publications and forums including print journals (Consciousness and Cognition, Journal of Consciousness Studies, Noetic Journal etc). The Center for Consciousness Studies at the University of Arizona in Tucson, USA, is the trail-blazer, with its large international semi-annual conferences since 1994. Parapsychology, Transpersonal Psychology, Consciousness and Experiential Psychology etc are common features in major psychology journals, since the 1990's. Studies on the relationship of consciousness to the humanities including theatre, fine arts, media arts and literature are also current. So far, however, there is not even a generally accepted definition of consciousness. Quantum physicist and Anglican Priest, Rev. Polkinghorne in "Exploring reality," (2005), posits a *dual aspect monism* that both mind and brain are made of the same *world-stuff.* They represent different aspects, or "complementary poles", of a single substance. The fundamental properties of the "mental pole" of this complementary duality interact with the rational, moral, and aesthetic qualities of human beings. "There is no real division between the natural and the supernatural. The universe is not a closed mechanistic system, and quantum chaos is a sign of ontological openness. Then God interacts naturally with the world by manipulating chaotic boundary conditions via the input of information" (p. 35- 36, 57-58). What perishes and what endures on death seems to be not an issue here or is murky.

The problem of sentience, sapience, intentionality and selfhood are all equally vexing and not amenable to analysis by ordinary (electromagnetic) tools. Subjective experience of **self** cannot be dismissed as "a quality of space time" (Psyche-D, 12 Feb 96). Experience is real. Space-time is an abstract mathematical concoction of coordinates, three real ones for space and one imaginary for time. It has nothing to do with experience! Different geometries represent different physical shapes, not varieties of experience, or qualia. Then there is quantum mechanical interpretations based on Eastern religions. Wave functions collapse in certain capillaries (microtubules) in the brain creating mental states (consciousness) which then collapses the wave functions and the circular reasoning goes on ad infinitum. Then there is *emergentism* where complexities of physical structures somehow cause mental states to emerge. Quantum and Relativity contributions are novel ideas, but there is nothing of any ontological import. On the other hand, quantum chemistry of ordinary "**light**-*matter*" and of extraordinary "**dark**-*matter*" may provide a scientific and physical basis for dualism and accentuate its explanatory powers for the problem of self-hood, by providing the physical link between the visible and the invisible domains. The two are intertwined, one permeating the other and coupled to it via *dark* and *light* chemical bonds and spin-spin interactions. It is incomprehensible to ordinary materialism. Bio dark-matter chemistry explicates what Maurice Merleau-Ponty in the "Visible and the Invisible" (page 133) calls "the flesh" an "anonymity innate to myself".

Intrinsic to all living matter is a phenomenon known as spontaneous emission of ultra-weak photons called biophotons. The source of biophotons remains elusive .Moreover, the emission rates per square centimeter per second differs vastly across the taxa, decreasing by an order of magnitude from plants to humans. This is not explained by any ordinary chemical process. Speculating DNA as the origin is indefensible on the basis of the verisimilar DNA structures and the very dissimilar biophoton emission rates across the taxa. Studies show that the emission rate in injured cells is higher than normal cells, and organisms with illnesses will likewise emit a brighter light, which has been interpreted as implying a sort of distress signal being given off. Injured cells have higher states of oxidative stress. Since no oxidative process is known to emit ultra-weak photons, attributing higher rates of emission of these photons in sick cells to higher oxidative strain is simply begging the question. Besides, the burst of biophotons at cell death cannot be logically explained on the basis of ordinary bioprocesses. More biochemistries in dying cells than in normal cells are unlikely. Chang et al (1998) studied this, but could not determine if this constitutes a "distress signal" or "oxidative stress". Gurwitsch (1922) thought this form of biophotonic signaling plays a role in the reception, transmission, and processing of electromagnetic data, primarily in the blood. The energetics and kinetics of biochemical processes justify neither the ultra-weak nature of these emissions nor the vast differences in the rates across the species. Neither Darwinism nor Punctuated Equilibrium has any provision or need for taxonomically dependent biophoton emission rates. The former is proven Natural Selection piggybacked with putative trans-speciation through DNA mutations. The latter was proposed by the late Prof. Stephen J. Gould within the framework of Marxism. He believed that Marxism with its sudden social change or revolution is the law of Nature. Darwinism requires gradual change. The fossil records show sudden change, instant transitions with no missing links. So he conjured up Punctuated Equilibrium. From an evidentialist perspective, both fall short of demonstrative, deductive validation. Punctuation is design and equilibrium is order, while Darwinism is chaos and chance. The spontaneous, continual and coherent bioiphoton emission and the species dependent emission rates point toward a primal order that is precisely and separately set for each species.

Defining a ratiocinative human life as Homo sapiens is blatantly confusing and incomplete as Prof. Marc Hauser (2008) states in his theory of "humaniqueness". Ordinary materialism will not account for such discrepancies, since there is no basic difference in particle configurations of electrons, protons and neutrons that separate humans from other animals or even plants. Bio dark-matter chemistry will provide answers to: **1.** Is there an invisible body permeating or supervening the visible body? **2.** Will it be structured and coupled to the visible body according to known physical laws? **3.** Will it continue to live after death of the visible body? **4.** What kind of particles constitutes a parallel invisible body? **5.** What is the source of biophoton emission? **6.** Why do biophoton emission rates differ by an order of magnitude from plants to humans? **7.** Is there a scientific basis for OBE/NDE, induced by trauma or by magnetic fields? **8.** Is physical resurrection possible? These are issues of deep interest in science, theology, philosophy and psychology.

Chapter 1
Ordinary Physicalism

Materialism attempts to identify intentional conscious acts with physical states. It ought to be redefined as *ordinary* materialism, if a parallel all-pervading dark-mater realm of particles also exists, with its own binding/bonding capacities and chemistries. *Ordinary* materialism can only conclude that, in the final analysis, our intellect is most probably encoded in electrons and photons at quantum level in a thermodynamic setting. Theoretically speaking, it takes only one electron to remember everything on a computer hard disk, if that information is phase-encoded in its fundamental modes of oscillation. Ahn, Weinacht and Bucks Baum (2000) from the University of Michigan were the first to demonstrate experimentally on a small scale, that information can indeed be phase-encoded in electrons using photons (light).The theoretical storage capacity of a single electron, or even a single photon of light, is phenomenal, even though the actual encoding and retrieving of the information using the technology is not available today (Meyer, D.A. et al 2000). A large number of backup copies can be made by encoding every electron, every photon and every other elementary natural oscillation. The discrete phase-encoding *spread* among many natural forms of oscillation over a wide frequency spectrum seems an attractive possibility. Ricciardi and Umezawa (1967) used quantum field theory to describe brain states, specifically the memory states they treated memory states as inequivalent representations of vacuum states of quantum fields, as in in the states of many-particle systems [10]. Stuart et al.(1978, 1979), Jibu and Yasue (1995) refined these proposals. Vitiello (1995), Pessa and Vitiello (2003) included effects of dissipation, chaos, and quantum noise. Vitiello (2001, 2002) give clear non-technical reports of current versions of this approach. For ordinary materialism ordinary matter, and energy are sufficient to explain all phenomena. It is a form of realism which assumes that a mind independent reality exists. Or, the world is really autonomous of the knowing subject, and everything that exists is material and self-caused. All phenomena, including those of the mind and consciousness, are ultimately reducible to the motions of ordinary matter. According to Moser & Trout, (1995) materialism is now a major ontology for philosophers and scientists. A materialist view is a holistic approach related to systems theory. Even the concept of information itself is understood from a materialist perspective (Hjørland, 2002; Karpatschof, 2000). Materialistic reductionism ignores all aspects of consciousness other than its input to behavior. Eliminativism considers thinking and feeling as mere theoretical or speculative concepts. A robot with exact human *behaviors* is human. There are no mental states and no intentionality. Mind is passé. It is a human concept with no hard scientific basis. Supervenience and emergence make the mind a neuronal epiphenomenon. The Identity Theory identifies the mind with complex neural structures. Stimuli cause brain states which lead to behavioral responses (Armstrong, 1968; Smart, 1963). It specifies mental states as dispositions. The materialist may accept the irreducibility of mind, but claim that its relationship to ordinary matter is more than mere causal dependence.

Actual particles of physics as the final reality constitute the basis of classical materialism. A philosophical pondering of the classical mind–body problem in the framework of quantum mechanics is known as *quantum mind-body* problem. Certain interpretations of quantum physics contradict this materialistic notion of an objective world existing independent of observation. The basic problem in quantum theory is this: If some system **S** is fully defined to be in state $\mathbf{S_0}$ at time $\mathbf{t_0}$, then what is the probability of specifying **S** to be in state $\mathbf{S_1}$, at time $\mathbf{t_1}$, immediately after $\mathbf{t_0}$? In Newtonian physics the predicted probability is unity or zero depending on whether $\mathbf{S_1 = S_0}$ or not. But quantum theory treats this as a solution to a statistical equation describing **S**, and gives an answer that usually is neither unity nor zero, but infinite odds in between. The key issue is the origin and significance of our subjectivity that chooses a preferred course of observation (measurement) of **S**. These subjective judgments are supposedly allowed by quantum theory to have the effects that they seem to make, and are attributed to a mental realm. Thus the theory is *interpreted* to make experience ontologically fundamental, while retaining intact all its mathematical and logical structures including statistical uncertainties. The wave function does *correspond* to a physical reality, but not *the* reality itself. It is an unproven and un-falsifiable presumption that consciousness causes collapse of a mathematical construct called wave function. The next assumption is that this very same collapse of wave function causes a moment of consciousness. That is circular reasoning. A series of these moments of collapse is supposed to form a stream of consciousness. Then, panpsychism automatically follows, because these processes are universally applicable to all particles and their supposed precursor waves. Pantheism and cosmic consciousness follow. This is mystical interpretation of quantum mechanics. Consciousness does not cause collapse of anything anymore than the name Katrina causes hurricanes!

Quantification of a property of matter by an equation, does not give that property itself any independent existence. Chance, chaos, entropy, temperature, volume of sound, intensity of light, form etc are quantifiable *qualities* pertaining to matter. But, none of these 'qualities' has an independent ontological existence. Real matter particles with mass behave **as if** in wave forms without mass. This specific *wave-like* property (not *wave* property) has no independent existence at any time, apart from the particulate matter at its fundamental level of existence. **There is no matter without mass and there is no wave with mass.** Since particles behave as if in wave forms, wave equations may be legitimately forced into and meaningful solutions arrived at. The absurd and the imaginary results must be discarded as usual. So much so, David Bohm (1990) theorized the simultaneous existence of both particle and its wave. Consciousness has as much to do with matter-waves, as the cock-crowing does for the dawn! These mystic notions constitute no testable theory. A thin gold foil can be easily put through a narrow slit, but the same foil when rolled up into a ball will not go through. It is not *consciousness* of the handler which makes the difference, but the measurable "real" malleability and ductility of gold that do it. These properties do not hover over gold as independent entities. The success of quantum mechanics is not because some consciousness-driven ubiquitous waves of matter float in space, but because matter has measurable wave like properties at its fundamental level of existence. Attributing a separate metaphysical existence to the invisible so called matter-waves is not warranted. In the brain, there are only fundamental particles of matter chemically and physically bound to form macro aggregates, no free drifting matter-waves to accommodate the notion of quantum mind or quantum consciousness.

If the hypothesis of *mind* or *consciousness* as a quantum physical product is true, then in the limit of a universal symmetry breaking at the psychophysically neutral level, every system must have both a mental and a material aspect. Since quantum phenomena are universal, everything must be mental and physical, be it animate or inanimate. Thus rocks, plants and animals must have equally demonstrable consciousness as in humans. Egalitarian panpsychism becomes the natural order (Skrbina, 2003). If so, how can panpsychism account for qualitative and quantitative differences between plant, animal, and human awareness, as well as the distinctions between different levels of human mental activities-conscious, self-conscious and unconscious? Awareness is not the same as self-awareness, the former is concerned with the environment, the latter involves awareness of awareness and requires a *physical* **self** to begin with.

Matter-waves and their superposition are just concepts of properties of particles, which become measurable at certain minimum dimensions of their real existence. But it is *matter* with mass which behaves *like* waves. This leads to befuddling measurement problems, since these behave both as particles and as waves depending on the method of observation! Consciousness has as much to do with it, as the cock-crowing does for the dawn! Attributing a separate metaphysical existence to the so called matter-waves (which are only wave-like properties of matter) is unreal. In the brain, various fundamental particles of matter are chemically and physically *bound* to form macro aggregates. There are no free, drifting matter-waves in neurons and synapses of the brain. The electromagnetic fields permeating neurons and synapses consist of real and virtual photons, in their wave states, and each being a disturbance in the photon field with no extra-physical mental properties. Even if they are mental, how can mind be stable and enable an immense density of information storage, in the constant flux of particles recycling in the cells at the atomic levels, sometimes within minutes? A durable non-electric dark matter body, a "mirror image" of the visible body, seems to be the answer.

A living cell is a *totality* that cannot be fully and completely known by studying all its differentiating and integrating processes. It may reproduce itself, or be given birth solely by some other living cell. There is no spontaneous generation of sentience, nor can an artificial living cell be produced by integrating the information obtained from knowing the differentiating processes of a natural living cell. Unlike rebuilding a disassembled mechanical motor, a living cell cannot be rebuilt from its elementary parts. Neurons are not universal cells for information processing that gives rise to subjective feelings and self-awareness. Otherwise, all animals must be conscious as humans. Especially so, since all animals from bacteria to humans possess the same or closely similar genetic molecular structures as well as control and regulatory mechanisms for their organ-functions including the brain. In humans, there are no quantitative correlations of mental capacities and performances with the number of neurons (size of brain), if they are required to produce subjective conscious experience. Consciousness is a unity. The brain is not one complex computer made of simpler (differentiated) computers/neurons organized (integrated) into a single network system. Every neuron is a holistic *totality,* and so also is the neural network. It is one whole *totality of totalities,* each with its own differentiations and integrations. Beck and Eccles (1990, 1992) and Beck (2001) give precise mechanisms of information transfer at the synaptic cleft. Even if a spectrum of consciousness exists across all complexities, it still offers no explanation of how these unconscious chaotic elemental-waves will or can produce the wholesome consciousness we enjoy. Why should neural correlates correlate into consciousness? Why should processes process into thoughts? Why should supervenience or emergence be conscious? Self-consciousness requires another layer of these same processes. To make sense, the universe itself must be alive to begin with, a notion cherished by the Vedas and shamanism and reckoned as postmodern science. It cannot be ruled out that ritual use of mind altering natural products widely known from primeval times may have a role in developing these notions. Today they are the trendy mind expanding or psychedelic drugs, presumed to be a fast means to acquire knowledge power.

Natural science claims that most, if not all, phenomena are intelligible in material terms. Mind is an exception. Consciousness or subjective experience, and the intentionality of thought which is the property of being about something beyond itself (*aboutness*), seem not to be physical in the *ordinary* sense. At the fundamental level of existence, Science depicts the world as made of quantum units of space-time which is the abode of everything, matter-waves which are wave-like energy-packets, and force-energy which assigns matter-waves their place in space-time. This is the *quantum trinity*. It is also believed that the space is filled with dark energy, and Higgs bosons which give to massive particles their mass and thus gravity. In this material milieu exists a tri-level sentience of plants, animals and humans. Science has to reckon Human **self** as an invisible but very active *stuff* of human life, possibly the locus of **mind**, an integral part of a triune system of sensory somatic apparatus (sensorium), intelligence machine (mind) and consciousness mechanism (awareness/self-awareness), a holistic shell of a dynamic *loop within loop*.

Quantum theory may play a role in explaining the relationship of consciousness to brain (Eccles, 1986). The patterns of electrical activity in brain create a model of a computer-self in the world of our existence. But, why life exists at all? How is it ostensibly an inevitable property of matter? If dark matter is included, possibly bio dark-matter chemistry may elucidate some of these mysteries, without mystical misapplications of quantum theory. It may resolve into a 'dark and light' problem. An invisible human body, a doppelganger, coupled to the visible human body is a reality.

Chapter 2
Bonding and Interaction of Particles

Why do matter particles interact at all? It is still a fiddly question that besets science. The puzzle is that things interact without touching! At the macro level, there is action (gravity) at a distance, even galaxies apart, without contact. The sun attracts the earth and the other planets. Two magnets (substances with a certain precise alignments of electron configurations) attract or repel as if they "sense" each other's presence. All that is available here is substance. Calling these forces as "gravity" and "magnetism" is begging the question. Gravity does not vanish when mass turns to energy. In fact, gravitation does not depend directly on mass; it acts through the energy (and momentum) of the matter involved. If the matter changes in some way to reduce its rest mass, this would have no long range effect on the gravitational field, because there is no net change in energy of the gravitating matter. It is generally the total energy (rest energy plus kinetic energy etc.) that is important for gravity, whether the particle in question has a rest mass or not. Gravity bends light rays (beams of photons). It acts on everything, including space itself! What causes it is simply unknown. Quantum gravity is yet another mystery. So is dark energy.

Physics does not know what these forces are. It can only say that at the fundamental level, a force is a thing which is passed between two particles. Winkler, K. et al. (2006), report that atom pairs can also bind together even through repulsive, rather than attractive, forces. This basically alters the molecular concept. Intuitively, a bound system of atoms implies an attractive force between them. However, to reach an equilibrium configuration, a repulsive force must balance the attraction. Otherwise the system would easily collapse. In the quantum world, attraction is not even needed to form a stable bound system. Repulsion can also affect a stable binding in the presence of a periodic spatial perturbation, where the energy cannot vary continuously, but is restricted to specific values. In order to conserve energy, repulsion between a pair of particles can lead to stable binding, if nonbinding would lead the two isolated atoms to have energies in a forbidden range (Bloch 2005, Feshbach 1958 and Fallani et al. 2006).

Substance reality of the ordinary materialism is a continuum of relational interactions. Chemical Bond is crucial for formation of complex bimolecular and all entities beyond the atomic structure. At the micro level, protons are held together with enormously strong nuclear forces within the submicroscopic nucleus, though like charges must repel on contact. Indeed they do repel outside the nuclear potential well. At the atomic and molecular level, electrons are held together in pairs in their orbitals around the nucleus, though we expect them to strongly repel each other. Electric charges alone cannot account for chemical bonding, stable electron configurations (periodicity of elements) is also a key factor. Mendeleev's chemical periodicity of elements, a law of nature, is not defined by a simple numerical relationship, or nomological formulae. Chemists attribute stability of atoms and molecules to duet and octet electron configurations, extrapolated from the empirical observation that the noble gases (rare gases) of the last group of the periodic table are the most stable elements. The number of electrons in their outer shells, 2 for helium, and 8 for the rest, complete the shells. Generally, atoms with incomplete outer electron shells are reactive. They contribute or share electrons between them resulting in the stable duet or octet configurations of the noble gases.

Quantum Mechanics is not in conflict with the chemical model of a molecule as a rigid structure of nuclei held together by shared electrons. The electron shells are only the quantum *energy levels* of electrons within each atom. The distribution of electrons within an energy level forms the electron clouds, or the *atomic orbitals*. An atomic orbital has a determined form and orientation and can maximally contain two electrons, each in a different spin state, up or down. These two electron waves 'mix' in a way that yields the lowest energy. The atom itself strives for a totally saturated outer octet or duet shell, because this is a very stable state (inert Noble Gas configurations). Quantum chemistry gives an almost full account of how atoms combine into molecules (and molecular ions). The theorem is only an approximation due to special quantum effects, particularly in excited states, for instance vibrational coupling that involves transitions between two structures. Energy states and electron configuration determine the way atoms can form bonds with each other resulting in molecular structures. In reality, it is spatial, temporal and spin relationships between electrons, each of which is unaware of itself and the relationships with each other. Dark matter also may have analogous affinities. Ackerman et al (2009) and Feng et al (2008) posited that "dark matter couples to a new long-range force known as "dark electromagnetism," mediated by particles known as "dark photons." Ordinary chemical bonding occurs when two nuclei simultaneously attract one or more electrons. A two-state system of **dark-matter atoms** may be set up qualitatively with energy levels split between them, which result in an energy bond. A **dark nucleus** may consist of positive monopoles (***psychons***) and ***neutrinos***. Adding a negative monopole (***emoton***) at a certain distance from the dark nucleus will constitute a dark atom. $\mathbf{E_0}$ is the ground state energy of two such nuclei **A** and **B** separated by an infinite distance. The ***emoton*** will switch positions between **A** and **B.** The exchange effects will split $\mathbf{E_0}$ into two sub-levels $\mathbf{E_0 + \alpha}$ $\mathbf{E_0 - \alpha}$ where $\boldsymbol{\alpha}$ is the probability amplitude for the exchange. The lowest energy level possible is when the ***emoton*** spends an equal time between **A** and **B**. It will depend on the distance between A and B. This causes "*bonding*". It is indeed much more complex than that, when various other *light* and *dark* particles are involved, but bonding energies and distances between **A** and **B** may be estimated for a qualitative basis of dark-matter chemistry. It has been proposed that they may yield dark Chemistries following the pattern of stable duet or octet configuration rules of ordinary chemistry, spin-spin interactions and couplings (Benjamin, 2003, 2007-A, 2007-B, 2010 A & B,

2012). Under suitable external conditions, a molecular dynamic system consisting of two or more element-systems results in the creation of a fundamentally real and stable new unit of physicality, a new totality. For example, the dynamic natural law governing two atoms at certain energy levels can be explained by quantum chemical determination of the energy states of atoms and molecules. According to this law, every free system tends toward the highest order of stability by striving to reach the lowest possible energy state. If two negatively charged electrons from the visible matter can reach a configurationally stable combination in an atomic or molecular orbital, it is not beyond the realm of reason to postulate a similar atomic and molecular configuration for axions of the dark matter, via possibly gravitons and spins.

Axions may be as diverse as ordinary atoms, with particles parallel to electron, proton and neutron. These may be respectively called Axion-e, Axion-p and Axion-n, with three different masses and Axion-e being the smallest. Like the electron, Axion-e may be a universal constituent present in plants, animals and humans, and conjugated to electrons. Axion-p bind to protons, but by genetic laws is present only in animals and humans, not in plants. Axion-n bind to neutrons and may be present only in humans. So, in humans the three Axions may form structures parallel to ordinary atoms. Perhaps gravitons or gravitinos may be involved, both negative and positive. Spin is another factor. These may have similar roles in axion atoms, as charge and spin in ordinary atoms. Quantum numbers abound in atomic theory. They represent quantum possibilities of energy levels that emerge, and could be applicable to the dark matter also. As in ordinary matter, dark-atoms will have the *principal* quantum number n, which can take the values 1, 2, 3, 4, 5, 6, ...n . Each **n** will have a principal energy level or a dark shell. Each shell will have n subshells. Each subshell has one or more orbitals within it; the **s, p, d, f, g** subshells have **1, 3, 5, 7, 9** orbitals respectively. The rules of placing axion-e within dark shells with nuclei made of axion-p and axion-n, may form the Dark Matter Aufbau principle and dark chemistries, yielding dark molecules, dark cells and dark bodies. Akin to the lighter ordinary atomic species, the lower Axion atoms, may have two main orbitals: the spherical orbitals or the s-orbital of the lowest energy levels (with the axion nucleus in the center), the dumbbell-shaped orbitals or p-orbitals of next higher energy levels (with the axion nucleus in the 'waist'). The rules of molecular structure, the directed valence bonds for each reactive atomic element, are natural laws governing conversion of simpler components into larger complexities. They could be applicable to axions also. The major difference is, while the wave-like properties of ordinary atoms or molecules can be ignored for most practical purposes, they are significant for all dark atoms and molecules, since they have relatively negligible masses.

Mental states and material states will be both physical, distinguished only as extraordinary and ordinary material phenomena. The brain is then a many-particle system of neurons permeated by their axion counterparts, at an energy level which directly correlates with mental activity. Mental activity is co-produced by the visible brain and the invisible axion counterpart. It is more likely that the entire visible body, permeated by an invisible body of axions, is involved in consciousness. Self-awareness could be a product of mutually recognizing resonant structures of *light* and *dark* matter. The mental and material manifestations may inherit mutual correlations due to the fact that they are jointly caused by the visible body and the permeating invisible body, con-centering into a holistic hypostasis with the possibility of a direct causal interaction between mental and material domains.

Chapter 3
Extraordinary Physicalism

Our present understanding of *ordinary* physical reality requires mind to be non-physical. Eccles rejected Descartes' notion of mind as a substance. How else can the **self** control its brain? He suggested a dualistic interaction system that relates to the two co-existent and parallel worlds of Popper (1959, 1977, 1995), where each is independent of the other for its existence and functions. Eccles anchored **self** to brain states, specifically the physiology of ultra-micro synapses. If each of these two worlds are made of ordinary matter and dark matter respectively, a theory of visible and invisible material systems that are ontologically independent of each other can be developed. The incorporation of the *extraordinary* physicality of dark matter components corresponding to electrons, protons and neutrons, into the living biosphere of the Newtonian world, will constitute an interiority that invisibly and immaterially permeate everything visible and material. This interiority counters Bohm's active information of the pilot wave as seat of consciousness. With a dark matter Aufbau in place, reality can be approached with a rich, logical frame work of ontology. Metaphysical abstractions can be replaced with actual physical theories. Physical *being* can be restored to its ontological status. Physicalism which is often synonymous with Materialism will then be fully meaningful. Ontology will be rescued from metaphysics into a physical realism. Creativity, awareness and self-awareness will be attributed to extraordinary natural elements permeating those that sustain determinism and the static. Heraclitus is right to state that "one cannot step twice into the same river", only if dark matter is excluded. Or else, it is only a half-truth. A non-electric permanent dark body, coupled parallel to the visible one, can become the stable repository of all events including thoughts. This will offer a consistent formal framework toward a novel dual-aspect physical scheme based on the distinction between tenseless axion time and tensed ordinary time. Dark matter could be the missing link to a fundamental descriptor of reality.

The shortest distance now measurable in physics is of the order of 10^{-16} cm, and Planck's unit is of the order of 10^{-33} cm which is the shortest possible length where current concepts of space-time are relevant. The range of scale in between is so vast that that it can accommodate almost limitless number of undetected structures including Bohm's Pilot Waves which could very well be the theoretical axions which lie toward the low end of that range, a tiny mass smaller than the electron mass divided by1 followed by 38 zeroes. Axions predicted by GUT are about 10^{-5} eV or $2x10^{-38}$ grams. [No axion particle has been detected so far [see http://xxx.lanl.gov/abs/astro-ph/9801286]. Bohm suggests that there is enough scale between the electron scale and the Planck scale that an electron can be very complex and possess a pilot wave. He attributes to electron an elementary sort of the will and consciousness to follow guidance, like an airplane following a beacon. This avant-garde notion is beyond the idea of Cooper-pairs in a superconductor which sense and steer around all the atoms without loss of energy, or Bose Einstein Condensates (BEC) adjusting instantly to environmental changes, and possessing non-locality. Father Jerome [http://go.to/QUFD] gives a metaphysical model that explicitly has consciousness existing in an axion BEC, replacing Bohm's pilot waves. However, he offers no means to couple the axions to the physical body. For that, bio dark-matter chemistry is needed.

Thought patterns may be waves in the axion condensate. Electrons merely follow their guidance. Parallel dark mater counterparts of the neuronal assemblies with negligible masses offer more effective wave-like properties. A dark matter body permeating the visible body, atom by atom, is less problematic and more pragmatic than Bohm's "pilot waves". Quantum theory will not rescue *ordinary* materialism from mysticism; extraordinary physicalism will. That involves dark matter and possibly dark chemistry. The parallel participation of both visible and invisible matter demonstrates a *real* supervenient relationship. Just as a chemical system (and less easily a biological system) can supervene on a physical system, a *dark-matter* body with encoded mental states can permeate and supervene the corresponding visible physical system. It is not *properties* of complex systems of ordinary matter that supervene, but one kind of *physicality* (non-electric and invisible) supervenes on another kind (electric and visible).

Dark-matter bodies, such as those made of axions, neutrinos, monopoles will have their own genes and genetic codes compatible with the ordinary genes. The human genetic codes on the dark genes recognize all the three axions (or bio dark-matter particles). These permeate the electrons, protons and neutrons, and are entangled with them. *Dark-matter* chemistries and structures are complete in humans with axion-atoms, axion molecules, axion cells and an axion body. The animal genetic code recognizes only axion-e and axion-p. *Dark-matter* chemistries and structures are less complete here. The plant genetic code recognizes only axion-e and therefore the *dark-matter* processes are least complete in plants. Though mental state parameters are represented in the dynamic inextricable maze of neural networks, Penfield (1975), the pioneer in brain surgery, after a lifetime of research insisted that the human mind definitely *does not* originate anywhere in the brain. Moreover, all cells including those of the neural networks are systems in flux at the molecular and atomic levels. For example, the synaptic membrane components are transient. In contrast, memory and cognition can endure, implying that mental states are encoded and hard-wired elsewhere on a more durable substrate. The most likely candidate for that are the non-electric *dark-matter* counterparts.

Information initiated by the release of transmitters in the presynaptic terminal, flow between neurons in chemical synapses (exocytosis). An incoming nerve impulse is rather too small to trigger this. So, thermodynamics or quantum mechanics is applied to statistically describe the trigger mechanism. "An effective mass below 10 electron masses is sufficient to ensure that quantum processes prevail over thermal processes", over a length of several nanometers, which is typical for biological micro sites (Quantum Approaches to Consciousness, Stanford Encyclopedia, May 2011. http://plato.stanford.edu/entries/qt-consciousness/). The trigger mechanism of Beck and Eccles (1992) is based on the quantum concept of quasi-particles which reflect the particle aspect of a collective mode. However, the indeterministic quantum trigger for stable mental states is questionable. Rather, the excitations of *dark* matter bodies provide the physical substrate for mental causation. The resonant vibrational excitations between the *light* and *dark* counterparts may offer a basis for self-awareness. Resonance is in essence recognition and response. It is counterproductive to invoke here Copenhagen interpretation of quantum theory and then in desperation conclude that "science cannot, by its very nature, present any answer to questions related to mind" (Beck 2001, pp 109f). Dark-matter chemistry gives an *extraordinary* physical dimension to solve at least some aspects of the mind-body problem.

Bio dark-matter particles may be of three different *flavors*: a **positive particle** of half-integer spin (**P**); a **neutral particle** of half integer spin (**N**); a **negative particle** of half-integer spin (**E**). In the biosphere, these three particles permeate and couple with protons (**p**), neutrons (**n**) and electrons (**e**) respectively. These couplings in the living biosphere are governed by genetic information code. The plant genetics allow only the coupling of **E** to electrons, the animal genetics permit an additional **P** coupling to protons, and the human genetics allow for the coupling of the additional **N** to neutrons. The three axions coupled to the human body will recombine to axion atoms corresponding to those with which they are entangled in the human body. No such recombination information is genetically encoded in plants and animals. There are no invisible rocks or trees or animals, because their basic structural units are not intrinsically coded to bind to *all* the three dark-matter particles. The invisible human body is uniquely made of the all the three dark particles.

For all we know now, mind and body are two mutually interacting entities. If mind is a mere form or property of matter, with no mass, no energy, it cannot interface with the body. Neuronal integration/differentiation of stimulus/response signals can only describe what brain *does*. The issue on hand is ontological, what **self** *is*. It cannot be just a bunch of bosons and fermions of ordinary matter. The phylogenic diversity with conspicuously different sensory, perceptual and cognitive abilities, in spite of verisimilar genes of living systems from the smallest microbe to the largest plants and vertebrates, is suggestive of different compositions of the substrate for different species. Why should the human mind be different from that of a plant or a beast which all have very similar or identical processes and structures including DNA? Why should it be different from the mind of a rock? The bosons/fermions are identical ubiquitously. There is a perceptible functional division between mind and body, while such division for a substance from its form or property is as unfeasible as separating sweetness from sugar. It would appear, what is needed is a real, physical, substrate of *mind matter* made of sub-atomic particles of negligible individual rest-mass. A non-electric substrate made of subatomic components of dark matter with appropriate spin and negligible mass is what constitutes self and provide the locus of mind. Autopoiesis of cells (Maturana & Varela, 1980) suggests an informed substrate, a common stuff of life, self-contained, self-referring unities coded for self-organization. These authors reason that the only reference living systems have is to themselves. The dynamic interactions of the visible and corresponding invisible structures coupled to each other physically may offer a possible understanding of the boundary between **self** and body. There is no requirement of a numinous dimension. Dark-matter chemistry will provide the coupling mechanisms via chemical bonds and spin-spin interactions.

Chemical properties of substances in general can be attributed to their electron configurations. However, that does not provide an adequate reason to attribute mental properties to electrons. Principles of quantum mechanics which exclusively deal with *conjugate pairs* of physical properties (momentum, position etc) cannot be causative factors of **self** or social phenomena. So also, Relativity should not be forced into Sociological and Psychological studies, without reference to the Absolute of Relativity. The velocity of light in vacuum is an Absolute. What are the Absolutes for Psycho-Social sciences? Who or what determines them? Biology can only speculate on how the ever- changing complex cellular and neuronal architectures and electrical transmission of nerve signals can convey messages or have coherent meanings and cognition. The constant flux within the cells at the atomic and molecular levels seems to imply that our very thoughts are mediated by an invisible physical realm. It is invisible in the sense that no one has ever managed to grasp a thought in a petri dish, as one might if it were an *ordinary* physical substance. Mental states are invisible. Yet they exist in the physical realm. They are coupled to a visible physical body. So these states must be physical. They cannot be a metaphysical abstraction, or an uncanny, non-material phantom that mysteriously operate within the body as some unknown essence or neural correlates that suddenly emerge or supervene as epiphenomena. Only physical *states* can emerge from or supervene the physical.

Plants are stimulus-response systems with a single level of sensory existence. Animals exhibit sensory qualities and sentience reflecting a bi-level order. In humans a triarchy of sensorium (sensory apparatus), sentience and sapience function harmoniously. Monistic materialism which look at the global unity of matter cannot fully account for such qualitative and quantitative differences across life forms made of the same matter. If this reflects a hidden extraordinary physicality, that will be a *bio dark-matter* realm with its own chemistry symmetrical to ordinary physical realm of matter made of electrons, protons and neutrons. It will be constituted of three different particles. Benjamin (2003) tentatively named these extraordinary particles as Emotons (**E**) parallel to electrons (e), Psychons (**P**) parallel to protons (**p**), Neumatons (**N**) parallel to neutrons (**n**), and monopoles (N **or +, S or -**) parallel to electric charges (+, -). Benjamin (2010, 2012) proposed an axion-like bio dark-matter and its chemistry involving spin-spin interactions. **Electron/emoton** pairs are common to all life forms,

proton/psychon will be in animals and humans only and **neutron/neumaton** will be unique to humans as shown below.

Plants = (e/E) + p + n	**Least complete**	**Most active**
Animals = (e/E) + (p/P) + n	**Less complete**	**Less active**
Humans = (e/E) + (p/ P) + (n/ N)	**Most complete**	**Least active**

The stability factors of these "dark bodies" will be in the ratio 3^1: 3^2:3^3, where the base 3 refers to the number of ordinary fermions **(e, p,n)** common to all, and the powers 1, 2, 3 refer to the respective number of axion like particles (**E, P,N**) in plants, animals and humans.

Species	**Particles**	**Stability Factor**
Humans	(**e/E**) + (**p/ P**) + (**n/ N**)	$\mathbf{3^3}$
Animals	(**e/E**) + (**p/P**) + n	$\mathbf{3^2}$
Plants	**(e/E) + p + n**	$\mathbf{3^1}$

Biophoton emission rates will be inversely proportional to the stability factor. This predicts an emission ratio of 9:1 in plants and humans, agreeing closely with experimental value of 10. Only humans will have the most complete set and most stable dark bodies. Differential incorporation of dark particles will be governed by both "dark" and "light" genetics, similar to XX and XY chromosome production. An additional scheme for the non-electric neutrinos will be the same, doubling stability factors without changing ratios of emission rates. Consequently, there is a **dark** physical substratum parallel to the **light** physical substratum. These two physical strata are in resonance with each other at all levels. Resonance between the visible and the invisible bodies coupled with **entanglements** is the basis for self-awareness, a **re-cognition** of each other. Resonance is recognition. his may also explain the taxonomic gradation of sentience. Plants have only sensitivity to the environment. Animals have intelligence in addition. Humans possess a third quality of self-awareness. From an ordinary physicalist view, no single basic difference in particle configurations, including about 1% disparity in about 20 genes, separates human beings from other animals or even plants. Panpsychism is inevitable for which there is no empirical evidence. Man is uniquely different from all other animals. Bio world needs bio dark-matter also. At death the dark body is left at a relatively negative energy state ($-E= mC^2$). This temporary dormancy may be quickened only by humongous external energy, leaving little chance for roving ghosts. Channeling, if it is true, is done by dark-matter entities existing at higher energy states. Dissociation of dark-matter chemical bonds from their corresponding "light" matter bonds in humans explains anomalous phenomena such as NDE/OBE. Low energy magnetic fields can also cause the same effect. Interactions between "light" and "dark" chemical bonds may be the source of biophotons. A differential distribution of three axion-like bio dark-matter particles gives stability ratios of dark matter bodies across the taxa that correspond to the experimentally observed ratio of biophoton emission rates in plants and humans.

Chapter 4
Bio Dark-Matter Body

Why should mind or consciousness exist at all? Can it occur in any and every type of material substance, or only in a specific kind of carbon-based molecular substrate? Are the properties of non-electric dark matter such as permeability, invisibility, suggestive of mentality? Where is the seat of consciousness? More specifically is its locus any part of the brain or is it endemic to the whole nervous system, or the whole body? What causes consciousness to oddly arise or emerge from matter or its structural forms and mechanisms? What is its function in the general scheme of mental activity and behavior of an organism? Who or what can be termed a conscious being? What function does phylogeny play in mental states of living entities? Is a plant, or an animal or a robot conscious as a human? A versatile robot must be human! But it is not.

There is a well publicized speculation that quantum mechanical effects in microtubules with pore diameter of about 14 nanometers cause the Platonic non-algorithmic (non-computable) elementary acts of consciousness (Penrose 1989, 1994, Secs 7.8 and 7.10; Hameroff and Penrose, 1996). The vast data from neurosciences throw no light on such conjectures. The "gaps in Penrose's toilings" are cited by Grush and Churchland (1995). Bohm (1990) presented smart electrons sensing active information of pilot waves as a real quantum mind stuff. Stapp (1999, p. 159) claims hybridizaion of "idea-like and matter-like qualities" of matter makes "two complementary and simultaneous modes of evolution".

Why particular regions of brain are active during particular mental activities is still unknown. Emanation of mental states from material brain states is not backed up by the results of any current neurobiology. There are conceptual ambiguities demanding clarification, concerning the continuous confusion of mental and material states (and their properties). If mental states are the primary objects of reference, the ordinary materialistic models show very little explicatory features of the subtle mind. Stipulations abound on mind-matter relationship ranging from the Cartesian (Descartes, 2000) duality of a fundamental distinction at an elemental level of existence to mind as an epiphenomenon of an extremely complex and highly ordered material system called the brain (Popper and Eccles, 1977, Eccles, 1994; Chalmers, 1996,). Quantum randomness is invoked because conscious free choices are fussy in a perfectly deterministic world. But randomness is not conducive to volition, and laws of statistics limit the quantum indeterminism of individual events. It is not reasonable to view mind-matter interactions in terms of causal relations in any theoretical frame work such as reduction, supervenience, emergence or quantum stipulations. (Kim, 1998; Stephan, 1999).

So far, all efforts to rescue ordinary materialism from the failure of reduction of all phenomena to elementary matter seem to be fruitless. It cannot opt for the supernatural or ontological metaphysics. Crafty schemes of epistemological autonomy go nowhere. The Ele/Mental problem relentlessly persists. How did elements become mental? If rocks develop any kind of consciousness, they will have concealed causal powers of mental states. Then, why is it that two rocks behave in an exemplarily consistent way all the time, and do not at least infrequently deviate from their consistent behavior? Complexity of organization has no intrinsic mental qualities and cannot be the elixir of phenomenal consciousness, since such complexities are universal and mind is not. Apparently, a universe devoid of a *mind stuff* is necessarily incapable of carrying any phenomenal consciousness. On its own, matter howsoever brilliantly draped in holism, emergence, supervenience, quantum mystery, chaos and ordered complexity, cannot transform into thought. The human body, specifically the brain, is a physical system that naturally manifests intelligence, without offering any adequate causal explanation for it. It is pitifully true that most of the present theories of mind are equally valid and valuable even if the cranium contained only a plastic sponge instead of the brain. *Ordinary matter* shows no promise of giving rise to mental property anywhere, or an interspecies or intra-species gradient in mental capabilities. An *extraordinary matter* as the *mind stuff* is a viable option.

It must be noted that no part of any living system is *materially* constant, except its structural form. The cells are constantly changing by growth, decay and replacement of its constituent fundamental particles. The chemistry of these structures is in a constant state of flux. The actin filaments in dendrites need replacing within 40 seconds (Star et al, 2002). Every five days, the NMDA receptors are replaced (Shimizu et al, 2000). Activity level controls post-synaptic organization and signaling via the ubiquitin-proteasome system. The entire post-synaptic density (PSD) - the protein packed zone that powers synaptic activity - is replaced, molecule for molecule, almost by the hour (Ehlers 2003). Myelin and RNA molecules are known to last months, and DNA is fairly stable, but they all need recurrent repairs and eventually all constituents particles are replaced regularly. This molecular chaos of ordinary matter raises the issue of unity, wholeness, and stability of consciousness and demands an extraordinary substrate. The whole brain is recycled constantly, about every other month, synapses being the most active. Only their shapes are maintained. Their component macro molecules disintegrate within minutes of their formation, as revealed by fluorescent tagging. They cannot maintain a steady identity. The new particles cannot carry the old information (memories), but systemic loss of memories or Kaleidoscopic personalities are not the norm. Evidently, the mind substrate must be different from any electromagnetic field associated with the molecules. A new kind of quantum physical substrate seems lurking in the dark with unique properties. What could be more appropriate than the extra ordinary non-electric dark matter?

The neuronal assembly activated by external stimuli makes the encoded content accessible to the dark counterparts. The activation leads to excited states with a finite lifetime and enables a conscious recollection of the content encoded in the axion ground state. Stuart et al. (1978, 1979) report detailed investigations of the stability of such states and the role of external stimuli. Vitiello (1995) describes a doubling of the collective modes of the system by its interaction with the environment. Axion counterparts of neurons in their natural states are not the classical thermal systems of charge carrying particles. They permeate the ordinary neurons and could be in highly ordered quantum states which are capable of coherent activity in neuronal assemblies. Coherent neuronal assemblies correlated to axion states could provide conscious states. Infinitely different coded axion states can result, offering the possibility of large thought processes and memory contents without overwriting. Ordinary matter with dark matter offers a viable physical dualism.

Dual-aspect approaches have a long history with beginnings in the distant antiquity. They view mental and material domains of reality as belonging to one underlying ontic reality of undifferentiated mind and matter, compared to the epistemological framework of the distinction between mind and matter. Quantum mechanical variations of this theme have been proposed by Jung and Pauli (1955); Meier (2001); Atmanspacher and Primas (1996) and by Bohm (1990); Bohm and Hiley (1993); Hiley (2001). Bohm and Hiley introduced implicate and explicate order to reflect the distinction between ontic and epistemic domains. Active information in implicate order informs the epistemic explicate order of mind and matter. The term information is used sometimes in the usual Shannon-type epistemic way and at other times as ontic and psychophysically neutral. The epistemic/ontic distinction refers to the distinction between a local reality of empirical facts obtained from classical measuring instruments and a holistic reality of entangled systems (Atmanspacher and Primas, 2003).

Pauli and Jung intuitively represented the distinction between epistemic and ontic domains of material reality as due to quantum phenomena connected by the process of measurement involving consciousness of observers. Jung's depth psychology view these two domains as connected by a process of emergence of conscious mental states from the unconscious, analogous to physical measurement, but includes a universal collective component consisting of the so-called archetypes. They constitute the psychophysically neutral level covering both the collective unconscious and the holistic reality of quantum theory. They order the psychical and physical manifestations in the epistemically distinguished domains of mind and matter, thus mixing epistemically dualistic and ontically monistic approaches. There is a causal relationship (in the sense of formal rather than efficient causation) between the psychophysically neutral, monistic level and the epistemically defined mental and material domains. Physicist Pauli and Psychologist Jung identify this kind of causation as the ordering operation of archetypes in the collective unconscious.

Relation between mind and matter is usually treated either at an epistemological or ontological level. Correlations are usually descriptive with empirical weight, while causation requires ontological theories to explain correlations. Causation entails correlations between cause and effect, but the converse is not always true. Correlations between two systems can have a common cause in their thread and not necessarily a direct causal relation. Basic sciences logically consider causal relations in terms of interactions. Physics recognizes four fundamental kinds of interactions (electromagnetic, weak, strong and gravitational) to explain the correlations in physical systems. However, mind-matter relation is much more complex, since empirical correlations between material and mental states now available are meager. These are not causal explications, only epistemological descriptions trying to salvage materialistic explanatory autonomy from the clutch of ontological reality. Science has a better alternative in terms of *extraordinary dark-matter.*

Three levels of sentience, plant, animal and human, exist in the in the biosphere. The plant sentience is solely somatic, functioning in emotons coupled to electrons (**E-e coupling**) of the plant cells. The animal sentience is somatic plus zoic with **E-e** and **P-p** couplings of animal cells. Human sentience is somatic, psychic and pneumatic with **E-e**, **P-p** and **N-n** couplings of human cells. Human body then is an invisible bio-system of dark matter atoms and molecules which are coupled to the light matter counterparts. A relational and holistic theory of human existence can be advanced based on an axion matrix in resonance with the ordinary fermion body. They interact with each other along the visible molecular reaction coordinates. Stimulus dependent transformation of visible processes of neuronal activities has their binaries in their dark-chemical counterparts in resonance with each other, with the possibility of one recognizing the other.

An invisible human body coupled parallel to the visible human body may exist. Plants having only one type of dark matter (corresponding to electrons only) and animals having had an additional type (corresponding to protons also) will not have completed dark-matter bodies. Stabilities of the dark bodies depend on the degree of completeness. Lesser the degree of completeness, greater the instabilities and larger the biophoton emissions rates. The increasing intensities of biophoton emission from human cells to plant cells as also the taxonomic differences in spite of the close DNA structural similarities support this view. Robots have no dark matter components since there is no DNA to govern these processes, and therefore no spontaneous biophoton emissions. No known chemical reaction including chemi/bio luminescence explains biophotons. The biosphere seems to have distinct photonic markers for its biodiversity which cannot be accounted for by the three fermions common to all matter. Besides, something more substantial than about 1% genetic difference must justify the difference between humans and chimps.

Only humans will have a complete hierarchy of dark bodies resulting in the invisible human body that exists parallel to the visible body. Since axions are non-electric, the axion body if decoupled can easily go through all material barriers. It will be durable and may not be governed by entropy Animals will have less complete, and plants the least complete form of dark bodies. Biophoton emission rates as a function of such stabilities seem to support this, plants emitting about 10 times more than humans. That is Extra Ordinary Materialism or Dark Matter Materialism. Quantum Theory, which has been highly successful in the physical realm, is considered to be incomplete. Perhaps it is not the theory itself, but the real constituents of the physical realm involved in this theory which are deficient. Coupled to electrons, protons and neutrons respectively, there could be three other physical, but non-luminous (non-detectable) dark particles in all living systems. Then, taxonomic differences including differential biophoton emission may be due to basic structural differences not observable by EM tools. Taxonomy may be related to biophoton emission rates with a predictable ratio of intensities for the different taxa.

The difference **between spiritual body** and **physical mind** seems more semantic than substantial. The resurrected body was a physical body which could be seen, touched and felt. Otherwise the Apostles must have been dreaming, hallucinating or delusional and unreal in the real world! The Pauline world view is not metaphysical, nor mystical, but real and observable (including the third heaven). Three thousand years before Paul, way ahead of Vedic mysticism, the same truth was revealed to Job: "When a man dies, shall he live again?.... yet in my *flesh* I shall see Him." There is no metaphysics here. A **neutrino body** and an **axion body** both permeating and coupled to each other and to the visible body will have enormous explanatory powers. The **inner-man = the neutrino body (psyche) + the axion body (pneuma).** The **outer-man = the body. That is** reminiscent of the Freudian **Id, ego, superego** constellation. Unlike the Freudian scheme, the processes here will be akin to the ***triune continuum paradigm*** of *spatiotemporal, constitution* and *information* continuums proposed by Naumenko (2002) cited also at http://research.triunecontinuum.com/tc-notion-logo.htm. It is a necessary triunity, because none of them exists spatiotemporally without the others. Unlike Naumenko's model, there is no abstract ontology or an abstract model here, only real physical structures genetically co-created at the moment of conception*. The visible body will be the *spatiotemporal continuum.* The subjective space-time metrics are used here for subjective representations. The axion body will be the *constitution continuum* which subjective constitutional metrics as subjective representations of objects in relation with their environs. The neutrino body is the *information continuum*. It relates information about objects and their setting to spatiotemporal intervals and thus contains information about mutual relations of the first two continuums.

The "**self**" thus is identical to the "**soma**" (the entire body exclusive of the germ cells) that **possesses it and is permeated by and integrally coupled to it.** Contrary to ordinary materialism, this inner world of **self**, or interiority, does not "arise" as an epiphenomenon out of the ordinary matter in the 'the learning, remembering brain encoded with representational maps.' The reality of interiority of the **self** cannot be different from the reality of the body itself. Else, **self** is a misnomer. **Self** cannot be any more potential than the body. The two are identical twin realities co-created at the very moment of conception. Both exist in the physical realm of matter, with this difference: one is of light-matter and the other of dark-matter.

***Note:** The question of virgin birth has been raised. Though beyond the scope of this book, bio dark-matter chemistry should be able to address it. A person to be 100% divine and 100% human will require that the 100% Divine must be *mighty enough* to become a 100% human zygote without any human agency other than being planted in a womb. Virginity will only certify authenticity of incarnation. The New Testament Trinitarian theology allows a rationale for the Almighty second Person of the Trinity to become the human zygote. A **unified field theory** or a **uniform field theory** in physics allows fundamental forces and elementary particles to be considered as terms of a single field. Field1 + Field2 + Field3 = 1Field. In terms of infinity, **Infinity1** + **Infinity2** + **Infinity3** = **1 Infinity**.

Chapter 5
What is Self?

The body is not the **self**; at the same time the **self cannot but** be **identical** to the **body.** Otherwise how can it be **self**? While both are equally **known,** the **self** is a **hidden** physical reality invisibly bound to and permeating the visible physical **body**. The former then is made of **non-electric** matter or **dark-matter** via its chemistry, in the same manner as the latter is of **electric** matter or **light-matter** via its chemistry. The **resonanc**e between the two may be the basis for **self-awareness**. Resonance is rudimentary recognition. How that resonance is transduced into awareness is not a chemical problem. It belongs to physics. **Niemark et al (2003) at Harvard present evidence for** "vibrissa resonance as a transduction mechanism for tactile encoding" in rats. *Bio Dark-matter chemistry* deals only with the "*what*" of the *objective* **self,** physics handles the *why* and *how and* neuroscience determines the *where and when* of "**self**" by locating the pivotal brain regions where the process of transduction occurs. The "**self**" thus is verisimilar to the "**body**" that **possesses it and is permeated by and integrally coupled to it.** Contrary to the perspective of **ordinary materialism**, this inner world of **self**, or *interiority*, does not "*arise*" as an epiphenomenon out of the **ordinary matter** in the 'the learning, remembering **brain** encoded with representational maps.' The ***reality of interiority*** of the **self** cannot be different from the ***reality of the body*** itself. Else, **self** is a misnomer. **Self** cannot be any more *potential* than the **body**. The two are identical twin **realities** *co-created* at the very moment of conception. Both exist in the physical realm of matter, the difference being one is of *light-matter* and the other of *dark-matter. Then* ***Spiritual body*** *is the same* as ***physical spirit***. That is extraordinary materialism

Number of the '*kinds*' of particles common for all categories of life is **3 (e,p,n).** In the biosphere, plants have a well-developed *one-dimensional* stimulus-response (**SR**) system. In other words, the **life dimension** (**LD**) of plants with respect to *dark matter* (**DM**) is 1. Only **e** and **E** are involved here. Animals have a strong *two-dimensional* (**SR + instincts**) or **SRI** scheme, with **LD** = 2. Here (**e** + **E**) is supplemented by (**p** + **P**) also. Humans have a *three-dimensional* (**SR + Instincts + Intuition**) or **SRII** ratiocinative format with **LD = 3,** where (**e** + **E**) + (**p** + **P**) is complemented with (**n** + **N**). The **LD** number thus correlates with the **DM** number. Dark matter itself has little or no interaction with light matter. The light and dark chemical bonds are susceptible to small changes in energies which cause release of photons. In reality, all living matter emits ultra-weak photons called **biophotons.** The emission rate depends on the position of the live organism on the taxonomy scale, plant cells emitting about ten times more photons per sq.cm per second than human cells (McTaggart, p.50, 2008). Stability factors based on complexity and diversity will justify this experimental observation. Matter can be divided as animate and inanimate with different **life dimensions** (**LD**) as follows: Inanimate (Robot): **LD = 0.** Plant **LD = 1**. Animals **LD = 2.** Humans **LD =3**. The combinatorial stability of the dynamic system of this **dark matter body** (**DMB**) in this diversity is exponentially proportional to complexity. Thus the stability ratio: Robot **DMB:** Plant **DMB** : Animal **DMB** : Human **DMB = 3^0 : 3^1 : 3^2 : 3^3 =1/3 : 1 : 3 : 9.** A value less than **1** means inanimate. The more the number of the "kinds" of dark matter, the greater the **DMB** stability, lesser their energy changes to produce any intrinsic photons. As discussed later, such photons uniquely exist in all living matter, plants are known to produce about **10** times more biophotons than humans which agrees closely with the formulation above. [Some concepts of this derivation are based on MacArthur, R.H in *Fluctuations of animal populations and a measure of community stability. Ecology 36, (1955) 533-536.* He suggested that the stability of a food web will be proportional to its complexity, a property which could be estimated by the logarithm of the number of links in the trophic web]. What is crucial then is not some fancy quantum mechanical (**QM**) *interpretation* to suit individual subjective predilections, but a uniform objective application of **QM theory** itself to equivalent *light* and *dark* physical situations. If the biosphere is made only of ordinary fermions (electrons protons and neutrons), then according to Prof. Johnjoe McFadden, Professor of Molecular genetics at the University of Surry, and one of the authors of Conscious Electromagnetic Information (CEMI) theory: "We have to accept that subjective experience is a component of the universe just as we have to accept gravity or mass as being features of the universe". Not so, if extraordinary particles are also included differentially in plants, animals and humans. It will also account for the ultra-weak biophoton emissions and the different emission rates across the taxa. The quantum chemistry of extraordinary particles such as axions, neutrinos, monopoles etc, will be just as valid as quantum chemistry of ordinary particles. An automobile may have mechanical life but not sentient life. Sentience may be mechanical, but not all mechanics is sentient. A crystal may mechanically grow and maintain itself by using materials from its environment. It has crystal life, not sentient life. In other words sentience may have some crystal life, but not all crystals are sentient. The **Deep Blue** may win all the chess games against its maker, **General Electric Co**, but will not know it has a maker/creator nor the pain of its creator losing all games against it, nor will it have any joy over thwarting its creator **GE**. Making DNA in the lab from scratch base materials will yield a dead DNA, unless the "scratch" itself has some life component in it. There are sapient and/or sentient life forms. Bringing into science *consciousness* as an undefined and indefinable metaphysical term is unwarranted. *Un-deadness* is a better term than consciousness which is just mist, myth and magic. The neurobiological conscious states are brain states which should not be conflated with the metaphysical *consciousness.* A thought analysis of self reveals the following:

1. One has an observable electromagnetic (**EM**) physical body.
2. One constantly communes with *oneself* or the "**I**".

3. This "**I**" has a very real but invisible existence in my visible physical body.

4. Then this "**I**" must also be invisible but physical.

5. Therefore the "**I**" must be a **non-EM** dark-matter physical body.

6. That suggests axion-like particles, neutrinos (three flavors), monopoles and their chemistries.

7. A triunity of a neutrino body within an axion-body within the fermion body is possible.

This "**I**" is invisible, but has got to be physical in order to reside in a physical body. . Physical interacts with the physical. Thus the very real "**I**" within the physical human body can be reckoned as an *extraordinarily* physical but invisible body made of invisible physical particles which will be non-electromagnetic, also known as known as dark-matter. First inference of dark matter came from the Dutch astronomer Jan Oort in 1932 studying stellar motions and then in 1932 by the Swiss physicist Fritz Zwicky at California Institute of Technology, CA, USA). The theoretical axions and neutrinos were suggested more recently as candidates for the cosmic dark-matter. However cosmological dark-matter has not been established as yet and there is considerable doubt today about its existence. A bio dark-matter realm may consist of axion-like particles, the three flavors of neutrinos and the monopoles. Its chemistries and *chemical bonds* should be able to produce this invisible body (or bodies) which will be coupled to each other via bonds involving particle spins and bond polarities. Bio dark-matter can stand on its own without any reference to astrophysical dark-matter.

These bonds are essentially electronic in nature for the visible **EM** body and magnetic for the invisible **non-EM** bodies. Though direct interaction between EM and non-EM particles per se will be negligible, *weak* energy changes involving EM and non-EM *systems* of chemical bonds could be sufficient for electromagnetic generation of *weak* photons. Indeed there is an intrinsic emission of photons which is persistent in all living cells day and night. This phenomenon is referred to by a variety of names, such as mitogenetic radiation, low level chemiluminescence, ultra-weak photon emission (**UPE**) or more commonly biophotons. Unlike the higher intensity chemiluminescence and bioluminescence, the biophotons are not easily visible without the aid of high resolution photomultiplier tubes. Progress in the development of low-noise, highly sensitive, cooled **CCD** (charge-coupled device) cameras which are able to count photons over thousands to millions of pixels enable imaging of biophoton emission in totally darkened chambers.

In all the reported anomalous events, whatever happens is seen or heard or felt physically. They disappear also physically. That smacks of a **non-electric** matter realm alternating between high and low energy states. The physical photons that reach one's eyes can come from a dark-matter physical object that may have a mechanism (or a dark-metabolism) to change its internal energies. Changes in energies can cause emission of photons. Even a mirage is an optical reality, only the water is the illusion, everything else is real. The water is a physical trick which is played on the eyes by the physical refraction of light rays in hot air over a physical surface.

An apparition need not be a departed soul of a diseased person on earth. The possibility exists that perhaps that a live being made of extraordinary matter can pretend to be a deceased person. I am constantly aware of my-*self* which is distinctly and discernibly different from my visible body. Experience is experiment where the experimenter is the experimentee. This invisible **self** exists, functions and permeates my visible electromagnetic (EM) body. Therefore the invisible **self** must be **non-EM**. I am a physical **EM-body** permeated and coupled to by an equivalent physical **non-EM** body with functions equivalent to or better in scope than the visible **EM body**. The resonance between the two may be a basis for self-awareness. Resonance is rudimentary recognition. When the visible body dies, the invisible but physical **non-EM** body will be decoupled (from the spin-spin interaction forces) and be deprived of an energy $E = mC^2$ where m is the body mass. Then the invisible but physical body will be dormant and in a relatively negative energy state unless energized into a different energy state by an external source. An energy change in a physical body may generate photons. As the whole body is involved, the whole body will produce photons in the shape of that body.

Biophoton emission rate shows an experimentally quantifiable taxonomic dependence; plant cells emit about 10 times more than human cells. Cells emit light with damaged cytoplasm, but not when the nuclei are removed or destroyed by ethidium bromide. According to Popp et al (1984) DNA is apparently at least an important source, since conformational changes induced with ethidium bromide in vivo are clearly reflected by changes of the photon emission of cells. However there is no report of the separated DNA itself emitting any light. Moreover, the DNA structures and processes are verisimilar across the taxa while the emission rates differ by an order of magnitude from plants to humans. They also concluded that the emission is *coherent* light, from the photon count statistics, the spectral distribution, the behavior of the emission after external illumination and from its passage through optically thick materials. Popp describes DNA as "exciplex laser system" collecting photons and emitting them as coherent light (exciplex is short for *excited complex*, also called excimer for *excited dimer*).

There needs to be a guiding principle with different dimensions to justify the existence of the different taxa with different degrees of sentience and sapience, both qualitatively and quantitatively. If the **principle** is a dynamic system of "**light**" and "**dark**" chemistries and chemical bonds, then **particles** and their **spins** come into play subject to quantum chemistry. At the chemical level, the DNA "double helix" are made from the same four chemical building blocks - called nucleotides. In that respect they are 100% same across the taxa. Only the sequence of arrangements differs. Protein formation has relevance to the sequence of nucleotides, not photosynthesis or light absorption or photon release. So there is no reason why biophoton emission rates/sq.cm/s differ by an order of magnitude from plants to humans. Moreover, what mechanism in DNA will make absorbed photons coherent when they are emitted? Identifying sources of variation that affect any process is as important as the process itself. In all engineering or physical or chemical processes there are two sources of variation: Common Cause and Special Cause. Common Cause variation is predictable and may be due to several factors. It is random and independent of each other. Special Cause variation is non-random, unpredictable and unstable over time. The variation of biophoton rates across the biosphere is non-random, but not unstable or unpredictable. So there is neither a **common cause** nor a **special cause**. Each species has its own independent cause of variation. Normal electromagnetic **causes** should produce normal electrical **effects.** Such is not the case here. A common external or internal **EM** source will not stimulate the verisimilar DNA of different species into different emission rates. There is no known in vitro photon production from DNA. There are reports of "burst of biophotons" when cells die and also sick or injured cells emitting more biophotons than healthy cells. One should keep in mind that biophoton emission is a cellular phenomenon. If it is an ordinary electrical effect in cells, then the photon intensities should be of ordinary magnitude. The rate of emission should not be vastly different- by an order of magnitude- from plants to humans.

Most materials give a transitory after-glow in the dark when previously exposed to light, but the after-glow emitted by all living cells and organisms is persistent. It is not as intense as fluorescence in certain chemical compounds, or the strong flashes of fireflies. Biophotons are spontaneously emitted with extremely low intensity. Even seeds emit biophotons. How much of that will be due to photosynthetic processes? There are specific molecules (or chromophores) that emit light of specific frequencies. However, biophotons are not of a single or a few frequencies as would be expected in the case of special light-emitting molecules. Instead there is a wide band, an unbroken range of frequencies. All evidences point to biophotons coming from the entire cell or organism acting as a unique *solid state* mechanism with energy stored throughout the organism. There is another effect known as delayed luminescence (**DL**). Here light is absorbed from an external source to excite the system as a whole, distribute the excited energy throughout the system and eventually to re-emit part of it as a wide band of frequencies. But this is a temporarily induced phenomenon of relatively higher intensity reflecting the complex excitation state of the whole. It is not intrinsic or spontaneous or continuous. The **DL** is an ultraweak reemission of absorbed photons after exposure to external monochromatic or white light. When the **DL** is over the constant tail of biophotons persists. Wang, Bókkon, Dai and Antal (2011) presented the first experimental proof of the existence of spontaneous ultraweak biophoton emission and visible light induced delayed ultraweak photon emission from in vitro freshly isolated rat's whole eye, lens, vitreous humor and retina. They obtained a photon signal of characteristic shape. There is a small decay region with a flux of photons changing by 2 to 3 orders of magnitude which is attributed to delayed luminescence. Then a long tail region remains with a constant emission of ultraweak biophotons. The decay character of the biophoton signal is not exponential, indicating a coherent signal. Their measurements show the coherent nature of biophotons emitted by different leaves at various temperatures in the range 15-50 degrees C.

Is there something about biogenic carbon itself that produce these biophotons? Tomato leaves have been studied for biophoton emissions. A tomato leaf has about 30-35% carbon by fresh weight and about 0.1% of fresh weight as chlorophyll (cultivated in the normal atmospheric concentration of CO2 which is about 0.03% by volume or 300 volumes of CO2 per million volumes of air, **vpm**). Plant cells have cell wall, chloroplasts, and a central vacuole in addition to all of the components of an animal cell. The mean carbon content of a human cell is about 18%. Granting all carbon atoms are engaged in some kind of bioluminescence the biophoton emission can be only twice more in plants than humans, but it is found to be ten times more (McTaggart, 2008). Since only 0.1% of fresh leaves is chlorophyll, it is all the more less likely to be the source of biophotons.

There seems to be an invisible **non-EM dark matter** realm as source for these emissions, such as **neutrinos**, **axions** and **monopoles**. There are many candidates suggested for dark matter such as axions, (Duffy et al,

2009, Erken et al, 2012), neutrinos (Davis et al, 1985, Lee et al, 2011) and monopoles (Gomez-Sanchez et al, 2011). Physicist Professor Nikolai Nijegorodov (2008) states: "in the nearest future physicists would prove, first theoretically and then experimentally that neutrino atom and neutrino molecules do exist. That would be enough evidence that life in the world of neutrino does exist". Okulov's (1981) neutrino model with magnetic charge (a point source of magnetic field analogous to the electric charge, ***e***) predicts a non-zero mass. Fermionic axions have been theoretically shown to exist (Hooper and Wang, 2004). Hu, et al, (2000) and Jerome (*http://go.to/QUFD*) indicate diversity of axions (*http://www.angelfire.com/ca/sanmateoissues/becdic/copacon.html*). Gravitons may behave as charged particles with negative and positive color charges as well as magnetic color (H. Javadia et al, 2009). Malkus (1951) determined that the binding energies of monopoles to matter are close to those of **chemical bonds,** that the Dirac magnetic **monopoles** interact with matter and could be separated by strong magnetic fields. If dark-matter is confined to the galaxies only and if it has no chemistry, it is of no routine use for the biosphere. Chemistry and chemical bonds are essentially "**spin**" phenomena. Dark matter chemistry, similar to SUSY chemistry (Clavelli et al, 2009) will involve monopoles to substitute for electric charges. Dark matter chemistry may give substance to the ancient notions of a "subtle body". The EM body will be the Maya. The notions of Forms by Plato and Aristotle could be explained in terms of dark-matter bodies. That is materialism extraordinaire.

The neutrino theory was first posited in1930 by Wolfgang Pauli as a solution to the missing energy in nuclear beta decay. Dirac's magnetic monopoles came up in 1931. The cosmic dark-matter theory is widely known since Jan Oort (1932) and Fritz Zwicky (1933) initiated the concept. More recently (1977) the theoretical axions were proposed by Roberto Peccei and Helen Quinn to plug a hole in the standard model. Axions, neutrinos and monoploes have been proposed as candidates for the cosmic dark-matter also. However, these *invisible* non-EM (*dark*) particles are useless for the biosphere if they simply whiz around as free particles. That is true of the *visible* (*light*) electromagnetic (EM) electrons (**e**), protons (**p**) and neutrons (**n**) also. The "*light*- particles" exist in the bound states yielding "*light*" chemistries and the **visible physical** bodies. Then the *dark*-particles may also exist in the bound states yielding "*dark*" chemistries and **invisible physical bodies**. As regards the biophotons, the taxonomic dependence of their emission rates/sq.cm/s could be correlated to a differential distribution of the *dark-particles* across the taxa. Humans have dark particles corresponding to all the three light particles, **e, p** & **n**. Animals have them corresponding to **e** and **p.** Plants have only those corresponding to electrons. The *ligh*t molecules and *light* cells are permeated by their corresponding *dark components*. These combinations will be governed *by light* and *dark* genetic principles analogous to XX & XY rules for chromosome formations. A robot will have no genetics, no dark matter components and no biophotons. Biophotons cannot be by-products of cellular metabolism or regulated cellular radical/redox processes, as these are of normal energy transactions and can produce only photons of normal intensity.

It is well-documented in the literature that biophoton output correlates to physiological changes in an organism. Cells of all kinds which are in stress emit more biophotons than healthy cells. Beans cut into pieces emitted 2–3 times as much light as whole beans, indicating that mechanical injury causes more photon emission from plants (Boveris et al., 1983; Gu & Popp, 1992; Hideg & Bjorn, 1996; Iyozumi et al., 2001; Nagl & Popp, 1987; Slawinska et al., 1992; Slawinski et al., 1992; Van Wijk et al., 1992) (Salin & Bridges, 1981; Slawinska et al., 1992) In the 1950s Strehler and Arnold (1951). Colli et al. were the first to present measurements of this spontaneous photon emission from a number of different plants (Colli & Facchini, 1954; Colli et al., 1955).

Cohen and Popp (1997) show that the biophoton emission in humans varies depending upon state of health, position on the body, time of day, and time of the year. Because of the nature of photomultiplier tubes (**PMTs**), images cannot be obtained. However, detailed images of biophoton emission can be obtained utilizing cooled, low-noise charge-coupled devices (CCDs) with exposure times on the order of minutes (Creath & Schwartz, 2004a). It is found that 89% of the emission from plant leaves is between 600–1000 nm putting it in the deep red and near infrared (IR) portion of the spectrum (detectors available today are not sensitive to emissions from the shorter wavelength ultraviolet peak). In plants, reactions relating to chlorophyll are known to be major contributors to emission in the red and near IR (Hideg et al., 1989). Experimental evidence points to chemical reactions involved in the production of singlet oxygen and other oxygen derivatives (free radicals) as the source of this emission (Slawinska, 1978). The work of Hideg and colleagues also points to similar reactions in mitochondria as another mechanism for emission peaking in the red and near IR present after many hours in darkness (Hideg, 1993; Hideg et al., 1991). Both chlorophyll and mitochondria are fundamental to cellular energy metabolism in plants. A common link between plants, animals and humans is that emission has been measured from isolated plant, animal and human mitochondria (Cercek & Cercek, 1979; Hideg, 1993; Hideg et al., 1991; Meduski et al., 1974). Mitochondria serve a vital role in the healthy functioning of cells and energy metabolism. There is also a correlation between singlet oxygen production (not related to chlorophyll or mitrochondria) and increased photon emission in plant, animal and human cells (Slawinska, 1978; Slawinski et al., 1978; Tilbury & Quickenden, 1988; Voeikov et al., 1999). Recent research indicates that the presence of free radicals, reactive oxygen species and singlet oxygen is associated with weakened immune function and various diseases (Ames, 2003; Bernardi et al., 1999; Chinnery & Schon, 2003). An advantage of developing a biophoton measurement model is that studies beginning with plants can be easily extended to animals and humans.

If the *human mind* is purely a *quantum* byproduct of ordinary particles, then pan-psychism is inevitable, for which there is no experimental evidence. Jeremy Taylor (2009) argues in *Not a Chimp* that the gap between humans and chimpanzees is far greater than molecular biology can indicate. The *humaniqueness* idea of Hauser (2008) seems obvious as opposed to the late Prof. Gould's notion of equating man with a twig since both are just a bundle of fermions. According to the tabloid newspaper *Sun* in UK, a record 227 paranormal events have surprised Britain in 2010, compared to the normal 140 to 180 reports per year. That number is rising steadily. *http://www.thesun.co.uk/sol/homepage/features/3153950/227-paranormal-events-last-year.html.* The Fringe Weird Report is compiled by the Rev Lionel Fanthorpe, a Cardiff-based Anglican priest and president of the Association For The Scientific Study Of Anomalous Phenomena. Ghosts are the most common phenomena, followed by UFOs, strange sounds and road phantoms, poltergeists, crop circles and shucks (specters of large black dogs). Intellectual snobbery cannot easily ignore it without betraying a gnawing sense of inadequacy or helplessness. Spookiness and sanctimony that often goes with these experiences will not lead anywhere. Physical interacts with the physical, call it by any name-spiritual, transcendental, ethereal, subtle, occult etc. Even a mirage is an optical reality, only the water in that reality is the illusion, everything else is real. The water is a physical trick which is played on one's physical eyes by the physical refraction of light rays in hot air over a physical surface. Similarly the physical "photons" that reach the eyes in these anomalous events must have come from live physical *dark* entities that interact with the physical environment. If these *dark* entities have a mechanism (or a kind of metabolism) to change its internal energies, they may cause emission or absorption of photons depending on the nature of the changes. In the case of emission, as the whole body is involved, the whole body will produce photons resulting in a visible radiant body.

Could these be the departed souls of deceased persons on earth? Or is it possible that perhaps a dark-matter entity is pretending to be a deceased person? One is constantly aware or conscious of one-***self*** which is distinctly and discernibly different from one's visible body. Since one is mentally experiencing this invisible **self** in one's visible electromagnetic (**EM**) body, the invisible **self** must be made of **non-EM matter**. One's physical **EM-**

Body is coupled to and permeated by a corresponding physical **non-EM** body with functions equivalent to or better in scope than the visible **EM** body. When the visible person dies (could be any moment) the invisible but physical **non-EM** person will be decoupled (from the spin-spin interaction forces) and be deprived of an energy $\mathbf{E = mC^2}$ where **m** is the body mass. In that sense, the invisible but physical person will be dormant in a relatively negative energy state of dormancy, unless energized into a higher energy state by a humongous external power.

Chapter 6
Physical Resurrection

"For the hour is coming in which all who are in the graves will hear His voice and come forth" (John 5:28-29). That will be a joyous reunion of family and friends. For as in Adam all die, even so in Christ all shall be made alive," declared the apostle Paul (1 Corinthians 15:22) "Behold, I tell you a mystery: We shall not all sleep [die], but we shall all be changed—in a moment, in the twinkling of an eye, at the last trumpet. For the trumpet shall sound, and the dead will be raised incorruptible, and we shall be changed (1Cor 15: 51-52). First Corinthians 15 (the resurrection chapter), verse 44 contrasts natural bodies which are subject to death with resurrected "spiritual" bodies, which are eternal. "For I know that my redeemer lives, and that he shall stand at the latter day upon the earth: and though after my skin worms destroy this body, yet in my flesh shall I see God; whom I shall see for myself, and mine eyes shall behold, and not another; though my reins be consumed within me."—Job 19:25-27. "Your dead will live; Their corpses will rise. You who lie in the dust, awake and shout for joy, for your dew is as the dew of the dawn, and the earth will give birth to the departed spirits" (Isaiah 26:19-21 "Many of those who sleep in the dust of the ground will awake" (Daniel 12: 1-3). Physician Luke declares that by "many infallible proofs" including His many appearances to the disciples after His death, Christ's resurrection is confirmed (Acts 1:3-5). He ate, walked and talked with them. They touched him and worshipped Him. Anglican bishop of Durham, Rev. N.T. Wright, one of the world's leading New Testament scholars speak of resurrection as "a new type of body living within a new type of world" (Jesus' Resurrection and Christian Origins, Gregorianum, 2002, 83/4, 615–635). His understanding is that "God will make a new type of material not subject to death out of the old material". That fits in with a description similar to a non-entropic, nonelectric dark matter body.

All ancient writings pertaining to soul including Plato and Aristotle seem to make sense in the light of bio dark-matter. The Vedic Maya makes sense if Maya is an EM physicality and dark matter the true reality. On Job 19: 23-27, Charles Haddon Spurgeon point out how the Patriarch was "impressed with a **sense of some weighty revelation concealed within his words" and** desired that it might be recorded in a book or and were graven with an iron pen and lead in the rock forever. Today two kinds of physicality are recognized: 1. Electromagnetic (EM) and 2. Non-EM physicality. The neuroscience has practically exhausted all possible EM "controls". Non-EM science is of very seminal beginnings now, not even widely accepted. The dark-matter science must be pulled away from astrophysics into biophysics, quantum chemistry, biochemistry and biology. It really does not matter what the astrophysicists are going to come up with. Reasonable arguments can be developed for **bio dark-matter** particles from **biophoton** data alone.

Hebrews 11:3 implies that the visible (matter) is made of invisible (fundamental) particles. There is no metaphysics implied there. Bodily resurrection is physical not mystical or metaphysical. In fact spirit and soul are derived from roots meaning the physical wind and breath. What is the difference between spiritual body and physical spirit? The resurrected body could be seen, touched and felt. The Apostles were not collectively dreaming or hallucinating. St. Paul claims that what we see now through a glass darkly will on resurrection be seen "face to face" and we shall know even as also we are known (1Cor 13:12). The Pauline world view is not metaphysical or mystical, but real and observable all the way to the third heaven. About three millennia him Patriarch Job said: "When a man dies, shall he live again?... Though after my skin worms destroy this body, yet in my flesh I shall see Him" (Job 14:14;19:26). Physical Resurrection is the foundational to Biblical faith. Eastern mysticism does not admit that. Physical means bodily, not metaphysical or mystical. If mysticism is real, the ultimate mystic must be someone risen from the dead. All recorded events of the post-resurrection deeds of the Messiah point to the contrary. He asked his metaphysically inclined doubters to touch him, eat with him and talk with him. Nothing is mystical here, everything is very physical.

A bio dark-matter body of negligible mass will meet all the requirements of a spiritual body. Monopoles, neutrinos, axions are dark-matter particles as diverse as ordinary particles. Electrons, protons and neutrons may have parallel 'dark' particles coupling with them via spin-spin interaction. Since Chemistry is in essence spin and configuration of electrons, it is conceivable that 'dark atoms' and 'dark chemistries' also exist. If so, a hierarchy of dark molecules, dark cells, dark organs and a 'dark body' may all exist coupled to their visible counterparts. Monopoles will be involved in that dark-matter chemistry.

Life here exists in an electromagnetic physical body in real space and time, not in hyper-space. Life hereafter will also exist in a body, a non-entropic, non-elctric, dark-matter body. Since astronomer Fritz Zwicky (1933), almost eight decades of observations and study has seen mounting evidence that dark matter accounts for the huge missing mass (about 90%) of the universe. Not only cosmology but particle physics also need dark matter particles, for example the theoretical axions, in order to plug a hole in the Standard Model. So then, what is death? It is a decoupling of the *dark-matter body* from the *light-matter* body. The visible entropic body dies and decays. The invisible non-entropic body may endure with all its memories and capacities fully intact. Unlike the entropic visible body, the invisible body cannot be destroyed by ordinary physical forces. The memories, feelings and subjectivities will continue to exist in tact. That does not necessarily mean that ghosts of the dead roam about. From a scientific point of view, when the visible body of say 150 lbs (energy beyond imagination) dies, the invisible body of negligible mass is left behind without that energy. That means it will be at a very negative energy state relative to the visible body. Raising it back to any functional state will require enormous energy of cosmic proportion to. It is very doubtful that any human activity will be able to provide that energy. As regards spirit beings roaming around, if at these reports are reliable and reproducible they will be constituted of nonelectric dark-matter and at a positive energy state adequate to interact efficiently with humans.

In Dr. Jack Kevorkian's words (caption on his painting **Nearer My God to Thee):** "This depicts how most human beings feel about dying — at least about their own deaths. Despite the solace of hypocritical religiosity and its seductive promise of an after-life of heavenly bliss death is the end and I am not afraid of the prospect of annihilation. I have accepted that there's nothing I can do to hold on to my life. Nothing will let me keep my life and I will have to die. I have accepted that much. Finally, as the novelist Anthony Burgess, says: the "vestigial fear of hell" is at most a conditioned reflex from a childhood indoctrination. We have no reasons for believing any of the claims that there is an afterlife." Life here exists in an entropic, electric body not in a hyper-space. If there is life hereafter, that will exist in a non-entropic, non-electric "dark-matter body". The invisible body made of axion-like Dark Matter is left dormant on death at a relatively negative energy state by the decoupling of the visible body ($-E = mC^2$)? Resurrection is possible, if energized from a vast external source. That is not metaphysics, may be extraordinary materialism.

What is **self** or selfhood? Is the "**I**" anything different from the physical body? If the answer is no, then neuroscience should have all the answers, but they do not have and in all likelihood they will not have. If the answer is yes then that cannot be **self,** rather *alien* or *intruder* will be more germane**. That is to acknowledge that one is disconnected from one's body.** Disparate symptoms will be prevalent globally. One's body will feel that something is disjointed or off. One's brain may conclude that there is an identity problem which needs be resolved and it will keep unraveling possibilities, but none of them will stay, since the **body** is disconnected from the **self**. If you believe your brain, your will go crazy! The dissonance will be incurable. Sanity or rationality is possible only if the **body** and the **self** are identical. The **self** and the **body** have to coherently run the show in concert. The only disparity allowable is that the **self** is invisible while the **body** is visible. The **self** is felt. The **body** is sensed. Sensing refers to the capability to process impressions generated physically through the sensations of sight, sound, touch, pressure, heat, cold, position and balance of our body. There is no struggle. Since no disparate symptoms or dissonance normally exist. The immediate question is why **self** is only mentally or internally or intuitively discernible and not physically detectable? Invoking mysticism or metaphysics is unwarranted from the scientific perspective. A bio dark-matter and its chemistry may possibly be the missing link between *dual aspect monism* and *Cartesian dualism.* The mental cannot be radically different in kind from the physical. Otherwise the two cannot be involved in any causal interaction. It must be noted here that the ordinary EM-matter and the extraordinary non-EM dark matter are both physical but of different category. The two bodies are identical, co-created from the moment of conception. The dark-matter body permeates the light-matter body and is coupled to it through chemical bonds and spin-spin interactions. Reliable first person narratives, including the occult, are reported which indicate physical contacts with invisible entities. These alien entities are in all likelihood constituted of invisible (non-EM) physical particles governed by their own genetics and quantum chemistries. The three flavors of neutrinos may have neutrino chemistry. The "human machine" will then be the most complex of all machines, a triunity of a neutrino-body within an axion-like body within the electromagnetic fermion body of electrons, protons and neutrons, three *totalities* in one. As already mentioned earlier, the dark bodies are incomplete in animals and plants.

The ontology of materialism rests upon the impression that the actual world including the biosphere can be explained physically in terms of the ordinary subatomic particles and their equations of state. Such extrapolations proved futile so far with respect to self-awareness, sentience and sapience. Perhaps, the omission of an *extraordinary* material realm of dark matter may be the key factor causing this impasse. Self-consciousness or awareness of awareness begins with the mysterious **self**. If neural correlates cause awareness, there must be neural correlates of neural correlates to cause self-awareness; likewise for supervenience and emergence. According to the ontological principle of the *identity of indiscernibles*, there cannot be separate objects or entities with all their properties in common. A mental state **M** must be a real *physical* state, if **it** were to couple to the physical state of the body. Since **M** is invisible, dark matter could be its likely physical realm. Physicality is the common denominator for both states, but the two realms belong to two different categories of matter. That calls for *extraordinary materialism*. Philosophy as the art of making distinctions between differences, must distinguish *ordinary materialism* based on ordinary matter from *extraordinary materialism* derived from extraordinary matter. That invalidates any materialistic egalitarianism across the inanimate and the animate worlds as the late paleontologist Prof. Stephan Gould affirmed. He claimed that there is no difference between a twig and a human being (both being made of the same material particles, electrons, protons and neutrons). The differential distribution of the dark matter particles across the taxa will support the notion of *humaniqueness* of psychologist Prof. Marc Hauser as mentioned earlier. The esoteric bio dark-matter and the ordinary matter are both physical. Bio-dark matter structures are non-electric and most likely non-entropic and durable. That has implications on life here and hereafter. At death the invisible dark-matter body remains either dormant at a relatively negative energy state ($-\mathbf{E} = \mathbf{mC}^2$), due to the loss of that much energy of the separated visible body of mass **m**. That can be resurrected by a vast external energy source. Dissociation of dark chemical bonds in optimum low energy magnetic fields explains Persinger's NDE and OBE experiments.

Conclusion

As for any enquiry, there are six questions pertaining to the problem of self-awareness: **who**, **what, where, when, why and how.** Neurosciences investigate the **where & when.** Physics explores the "**how** and reservedly **why**". Chemistry studies the *who* & **what.** Psychology, philosophy, theology and other social sciences deal with the **why**. The very notion of a mystical consciousness is outside the field of all true sciences. Much against the strong disapproval by Einstein, it was unnecessarily forced into quantum mechanics by the Copenhagen interpretation, rather *Copenpagan invocation.* There is no factor for mysticism anywhere in any of the **mathematical or statistical** equations. What is consciousness is not the real question. Conscious of what is a genuine question. **Self** is the immediate answer. **Self** cannot be the **body**. At the same time, it cannot be anything **totally** different. Otherwise how can it be **self**? **Self** and **body** must be nearly identical twins formed at the moment of conception. **Self** is not **mind,** most probably it is the locus of mental events, though often the two words are interchangeably used. Only, the former is invisible and the latter is visible. That **only** means one is made of invisible (non-electric) matter and the other of visible (electromagnetic) matter. In other words, s**elf** is made of **bio dark-matter.** The **body** is made of **light-matter. Bio Dark matter chemistry** is **symmetrical** to **bio light-matter chemistry**. What the "**electric poles**" are to ordinary chemistry will be what the "**monopoles**" are to extraordinary chemistry. A **neutrino body** and an **axion body** both permeating and coupled to each other and to the visible body will have enormous explanatory powers. The **inner-man = the neutrino body (psyche) + the axion body (pneuma).** The **outer-man = the body. That is** verisimilar to the **Id, ego, superego** constellation. The "**self**" thus is identical to the "**body**" that **possesses it and is permeated by and integrally coupled to it.** Contrary to **ordinary materialism**, this inner world of **self**, or *interiority*, does not "*arise*" as an epiphenomenon out of the **ordinary matter** in the 'the learning, remembering **brain** encoded with representational maps.' The ***reality of interiority*** of the **self** cannot be different from the ***reality of the body*** itself. There is no need for unnaturalness in sciences. Self-awareness may involve resonance between the light and dark matter bodies. Resonance is rudimentary recognition.

Dark matter processes are existentially fundamental just as those of the visible substance. It offers a dynamic picture of reality in which the basic unit-events in space-time are not vacuous, but can develop a subjective nature through various physical mechanisms such as non-locality, resonance, replication, storage, retrieval, entanglement etc. The physical processes allow attributes that might be labeled cognition, memory, creativity, awareness, and self-awareness. What Whitehead (1978) calls *prehension*, is parallel computing and resonance between the luminous and the non-luminous bodies. Life is characterized by the co-*creativities* of visible and invisible physical processes. Reality is a *resonant* dynamism of parallel processes. This is a genetically active, self-determining substrate that creates a subject-object distinction in a temporal dimension. The **self** begins at conception as a DNA-governed matter/dark matter resonant process. Experience emerges via resonance states that recognize each other. Recognition is awareness. This process is neuronally transduced into thoughts, subjectivities and memories that can be stored permanently and retrieved repeatedly from the non-electric, non-recycling parallel brain.

Complex objects exist as actual things that stimulate responses in the sensorium. Parallel events occur also in the corresponding non-luminous bodies. These occasions are processes which possess a defining subject-object unity, resulting from the resonance bonds between them. They are reciprocal 'actions' of the resonant constituents of the visible-invisible matrix. The entities in resonance act on each other communally, capable of modifying each other in some details by the entangled association. The unity between the entities of increasing orders of interrelated complexities is more like the summation of fields into a single field than that of a collective whole. It transcends the sum total of all the constituents, yet a single totality, a whole substance reality. The parallel dynamic dark matter body of axions explains the existential argument that we can only rightly understand the physical reality around us as a correlation with our own subjectivity that we invisibly experience, allowing a link between experience and consciousness.

Process thoughts, by and large, incorporate the views of nested or concentric elements of reality (Rescher 2000), similar to the ideas of Holon (Koestler, 1967) and Holoarchy (Wilbur, 2000). A dark matter body is a complete totality pervading another complete totality to which it is constantly coupled as a parallel constituent part, thus forming a dynamic relational set. Efficient causation alone is fundamentally insufficient to explain the phenomena of consciousness in ordinary matter. Relationships between ordinary constituents are crucial, but by themselves alone, not causal for consciousness. More importantly the parallel dark constituents enable coherence via closure of these sets of ordinary relationships. One set of temporary *electric* relationships in the visible substrate closes into a permanent-level set of non-electric relationships in the invisible substrate.

Science and technology reckon structure as the formal cause resulting from efficient causes. The reverse is often the case, i.e. formal causes produce efficient causes. Dark matter structures as formal cause of an event determine which of the many possibly efficient factors are actually effective. Qualia related subjective experience, not necessarily consciousness, could be purely a structural phenomenon involving certain axion components. As Whitehead states: "Consciousness presupposes experience, and not experience consciousness" (Whitehead 1978 [1929], p. 53). Consciousness is a higher order subjectivity, caused by processes of resonance and entanglement between two *totalities* in parallel, one dark and the other visible. Just as the visible is made of electrons, protons and neutrons, the invisible is made of axion-e, axion-p and axion-n. The dark body wholly permeates the visible and is coupled to it, forming a single whole of two wholes. Kant's noumena, things-in-themselves, as ultimate reality exist independent of perception and experience and are beyond detection. They are distinct from phenomena, products of subjectivity, sensible mental intuitions, the things-as-they-appear to the senses, including space and time. Constructivism argues that noumena do not exist and the universe is only phenomena (mental). For empiricist David Hume, the notion of noumena is meaningless metaphysics. All this makes Science good for nothing, except for some exciting feelings. The dark matter and its dark chemistry clarify these subject-object ambiguities. -

A plant has life with no known mind of any kind, unless panpsychism is forced into it. From what we know empirically, animals may have animal minds but no human minds. That is not justifiable, if plants, animals and humans are made of the same electrons, protons and neutrons. Animal and human brains are systems of same or similar complex structures and functions. The IQ of Mother Nature is practically zero, until the smallest sentient cell steps in with an instant basic minimum of intricacy for survival and with information that may exceed quantitatively that of the Library of Congress. Much work is done on consciousness. However, nobody can confirm a location of consciousness. It is all over as if permeating every cell, yet within the confines of a physical architecture of atoms. There is a level zero of no sentience in dead matter. There is a level 1 sentience in plants, level 2 sentience in animals and a level 3 sentience in humans. Whatever values may be assigned to these levels, there is an observable, measurable difference. But it is the same insentient particles that they are all made of, unless some unknown invisible particles X, Y, Z are also present. If Partcle-X is present in level 1 then Particle-Y is additionally present in level 2, and Particle-Z is present in level 3. The successive addition of an extra particle X, Y, Z to the successive levels explains the observed gradation of the sentience. Electrons, protons and neutrons have all their axion-like counterparts yielding the invisible body only in humans. Plants and animals have only incomplete shells of axion-like particles. Only in humans is there the possibility of a stable invisible body. The vast differences in biophoton emission rates across the taxa by an order of magnitude seem to support this gradation of stability.

Acknowledgments

Author acknowledges the immense help received from the scholars whose articles are cited and included in references of this manuscript and is grateful to authors / editors / publishers of all those articles, journals and books from where the literature for this article has been reviewed and discussed. The author gratefully appreciates acceptance of his earlier papers by editors/professors of different journals and proceedings.

References

Acharya, Bobby S., Gordon Kane, Piyush Kumar. Compactified String Theories -- Generic Predictions for Particle Physics. Int. J. Mod. Phys. A, Vol. 27 (2012) 1230012 Ackerman L, Buckley MR, Carroll SM, Kamionkowski M. Dark matter and dark radiation, Phys. Rev. D 2009;79 (02) 3519.

Ahn, J; T.C. Weinacht, and P.H. Bucksbaum, Quantum information storage and retrieval in Rydberg wave packets.Science 287, 463 (2000)

Armstrong, David M. 1968. A Materialist Theory of the Mind. New York: Humanities.

Atmanspacher, H., and Primas, H. (1996). The hidden side of Wolfgang Pauli. Journal of Consciousness Studies 3, 112-126. See also H. Atmanspacher, H. Primas, and E. Wertenschlag-Birkhäuser, eds (1995). Der Pauli-Jung-Dialog. Springer, Berlin.

Atmanspacher, H. (1997). Cartesian cut, Heisenberg cut, and the concept of complexity. World Futures 49, 333-355.

Atmanspacher, H., and Primas, H. (2003). Epistemic and ontic quantum realities. In Time, Quantum and Information, ed. by L. Castell and O. Ischebeck, Springer, Berlin, pp. 301-321.

Beck, F., and Eccles, J. (1992). Quantum aspects of brain activity and the role of consciousness. Proceedings of the National Academy of Sciences of the USA 89, 11357-11361.

Beck, F. (2001). Quantum brain dynamics and consciousness. In The Physical Nature of Consciousness, ed. by P. van Loocke, Benjamins, Amsterdam, pp. 83-116.

Benjamin, Philip. Bio Dark-Matter Chemistry", International Journal of Current Research and Reviews Vol 4 issue 20, 2012

Benjamin, Philip. Dark Chemistry & the Paranormal, WorldComp'10 Proceedings of the International Conference on Artificial Intelligence, Vol II, pp 633-39, July 12-15, 2010, CSREA Press, Las Vegas, NV.

Benjamin, Philip. Dark chemistry & paranormal phenomena, International Journal of Applied Science & Computations, Vol. 17, No. 1 Pages 16 to 36, June 2010

Benjamin, Philip. Dark Matter & Dark Chemistry NeuroQuantology 2007, Vol 5 # 3, 322-326; Dark Chemistry or Psychic Spin Pixel? NeuroQuantology, June 2007, Vol. 5 #2, 197-204.

Benjamin, Philip Mind Matter, Noetic Journal Vol 4 # 4, 351-360, 2003 [Nobelist Sir John Eccles Centennial Edition]. **http://biodarkmatter.webs.com/index.htm**

Bohm, David. (1990). A new theory of the relationship of mind and matter. Philosophical Psychology 3, No. 2, 271-286. Bohm, D., and Hiley, B.J. (1993). The Undivided Universe, Routledge, London. http://members.aol.com/Mszlazak/BOHM.html

Boveris, A., Varavsky, A. I., Goncalves da Silve, S., & Sanchez, R.A. (1983). Chemiluminescence of soybean seeds: Spectral analysis, temperature dependence and effect of inhibitors. Photochemistry and Photobiology, 38, 1 99–104.

Bloch, I, (2005) Nature Phys. 1, 23-30 (2005)

Chalko, Thomas J., NU Journal of Discovery Vol 2, March 2001, http://nujournal.net/, Natural Uni 2001, http://naturaluniversity.net/

Chalmers, D. (1996). The Conscious Mind. Oxford University Press, Oxford.

Chang, J.J., and F.A.Popp: "Biological Organization: A Possible Mechanism based on the Coherence of Biophotons". *Biophotons.* J.J.Chang, J.Fisch and F.A.Popp, eds.

Clavelli L, Sanjoy S. *Covalent Binding in a Susy Background.* Int. J. Mod. Phys. 2009; A24: 4245.

Coherence of Biophotons". In: *Biophotons* (J.J.Chang, J.Fisch and F.A.Popp, eds.), Kluwer Academic Publisher, Dordrecht-London 1998, pp. 217-227.

Cohen, S., & Popp, F. A. (1997). Biophoton emission of the human body. Journal of Photochemistry and Photobiology B-Biology, 40, 187–189.

Colli, L., & Facchini, U. (1954). Light emission by germinating plants. Il Nuovo Cimento, 12 150–153.

Colli, L., Facchini, U., Guidotti, G., Dungani Lonati, R., Orsegnigo, M., & Sommariva, O. (1955).Further measurements on the bioluminescence of seedlings. Experentia, 11, 479–481.

Creath, Katherine, and Gary E. Schwartz. (2005). What Biophoton Images of Plants Can Tell Us about Biofields and Healing. Journal of Scientific Exploration, Vol. 19, No. 4, pp. 531–550, 2005. http://www.scientificexploration.org/journal/jse_19_4 _creath.pdf

Descartes, René., *Philosophical Essays and Correspondence,* ed. by Roger Ariew (Hackett, 2000)

Davis M, Efstathiou G, Frenk, CS., & White SDM. "The evolution of large-scale structure in a universe dominated by cold dark matter". Astrophysical Journal 1985; May 15, **292**: 371–394.

Duffy DL, Karl van Bibber. Axions as dark matter particles. **New J. Phys.** 2009. **11:(**10) 5008.

Eccles, J. C., Proc. Roy. Soc. 227B, 411-428 (1986)

Eccles, J. C. (1990). A unitary hypothesis of mind-brain interaction in the cerebral cortex, Proc. Roy. Soc. London B 240, 433-451.

Eccles, J. C. (1994). How the Self Controls its Brain. Berlin: Springer-Verlag Ehlers, M.D., (2003), Nature Neurosci. 6:231-242.

Erken O, Sikivie P, Tam H, Yang Q. Axion Dark Matter and Cosmological Parameters. Phys. Rev. Lett. 2012; 108: 06 1304.

Fallani, L et al (2006) Feshbach, H., Ann. Phys. 5, 357 (1958)

Feng JL, Tu H, Yu, & Hai-Bo. Thermal Relics in Hidden Sectors. J. Cosmology & Astroparticle Physics 2008;10: 43.

Gell-Mann, Murray and Zweig, George 1964

Grush, R., and Churchland, P.S. (1995). Gaps in Penrose's toilings. Journal of Consciousness Studies 2(1), 10-29. Response by Penrose, R., and

Gu, Q., & Popp, F. A. (1992). Nonlinear response of biophoton emission to external perturbations. Experientia, 48, 1069–1082.

Gurwitsch, A.G., "Über Ursachen der Zellteilung". *Arch. Entw. Mech. Org.* 51 (1922), 383-415. http://www.quantumbionet.org/eng/index.php?pagina=52

Hameroff, S. (1995). Journal of Consciousness Studies 2(2), 98-111.

Hameroff, Stuart, (1996), Conscious Events as Orchestrated Space-time Selections, Journal of Consciousness Studies, 3(1):36-53 (1996)

Hameroff, Stuart & Rogr Penrose (1996) Orchestrated Reduction of Quantum Coherence in Brain Microtubules: A Model for Consciousness. Stuart Hameroff & Roger Penrose, Toward a Science of Consciousness - The First Tucson Discussions and Debates.

Hepp, K. (1999). Toward the demolition of a computational quantum brain. In Quantum Future, ed. by P. Blanchard and A. Jadczyk, Springer, Berlin, pp. 92-104

Hiley, B.J. (2001). Non-commutative geometry, the Bohm interpretation and the mind-matter relationship. In Computing Anticipatory Systems - CASYS 2000, ed. by D. Dubois, Springer, Berlin, pp. 77-88.

Hjørland, 2002 Hjørland, B. (2002), Epistemology and the Socio-Cognitive Perspective in Information Science. Journal of the American Society for Information Science and Technology, 53(4), 257-270.

Hideg, E., & Bjorn, L. O. (1996). Ultraweak light emission, free radicals, chilling and light sensitivity. Physiologia Plantarum, 98, 223–228.

Hu, Wayne, Rennan Barkana, Andrei Gruzinov, Cold & Fuzzy dark Matter Phys. Rev. Letters, 85, 1158 (2000) astro-ph/0003365

Iyozumi, H., Kato, K., & Makino, T. (2001). Spectral shift of ultraweak photon emission from sweet potato during a defense response. Photochemistry & Photobiology, 75, 322–325

Javadia H, Forouzbakhshb F, Pour Imanic H. A New definition of Graviton.General Science Journal. 2009; 1-11.

Jibu, M., and Yasue, K. (1995). Quantum Brain Dynamics and Consciousness, Benjamins, Amsterdam

Jung, C.G., and Pauli, W. (1955). The Interpretation of Nature and the Psyche. Pantheon, NY. Translated by P. Silz.

Kim, J. (1998). Mind in a Physical World. MIT Press, Cambridge Ma.

Karpatschof, B. (2000). Human activity. Contributions to the Anthropological Sciences from a Perspective of Activity Theory. Copenhagen: Dansk Psykologisk Forlag. ISBN: 87 7706 311 2

Koestler, Arthur in "The Ghost in the Machine" (1967) and was presented again at the Alpbach Symposium (1968) in a paper titled: "Beyond Atomism and Holism - the concept of the holon".Malekin and Yarrow, 1997 Malekin, Peter, and Ralph Yarrow. *Consciousness, Literature and Theatre: Theory and Beyond.* London: Macmillan, 1997.

Lee, Hye-Sung, Liu Z, Soni, A. Neutrino dark matter candidate in 4th generation scenarios. Phys. Lett. 2011; B704: 30-35. 23.

Malkus WV. The Interaction of the Dirac Magnetic Monopole with Matter. Physical Review 1951; 83 (5), 899-905.

Maturana, H.R., Varela F.J. Autopoiesis and Cognition: The Realization of the Living (Boston Studies in the Philosophy of Science, Vol. 42) Reidel, Holland 1980.

McTaggart Lynne. The Field, p. 50, 2008. Harper. ISBN 978-0-06-143518-8 (pbk).

Meier, C.A., ed. (2001). Atom and Archetype: The Pauli/Jung Letters 1932-1958. Princeton University Press, Princeton. Translated by D. Roscoe. German original Wolfgang Pauli und C.G. Jung: ein Briefwechsel. Springer, Berlin,1992.

Merleau-Ponty, Maurice. The Visible and the Invisible. Evanston, Illinois: Northwestern University Press, 1968.

Metzinger, T., (2003), pp. 370, 385, 390. Being no one. Cambridge, MA: MIT Press.

Meyer,D.A., Kwiat,P.G, Hughes,R.J., Bucksbaum,P.H., Ahn,J., Weinacht,T.C., Does Rydberg State Manipulation Equal Quantum Computation?. *Science,* 289, 1431 (2000).

Moser & Trout, (1995) Moser, P. K. & Trout, J. D. (1995). Contemporary Materialism. A Reader. London & New York: Routledge.

Nagl, W. & Popp, F. A. (1987). Opposite long-range interactions between normal and malignant cells. In Barrett, T. W., & Pohl, H. A. (Eds.), Energy Transfer Dynamics (pp. 248–256). Berlin, Germany: Springer-Verlag

Naumenko, A. "Triune Continuum Paradigm: a paradigm for General System Modeling and its applications for UML and RM-ODP", Ph.D thesis number 2581, Swiss Federal Institute of Technology - Lausanne. EPFL, June 2002.

Neimark, Maria A. Mark L. Andermann, John J. Hopfield,3 and Christopher I. Moore. Vibrissa Resonance as a Transduction Mechanism for Tactile Encoding. The Journal of Neuroscience, July 23, 2003,23(16):6499–6509 •6499

Nijegorodov N. "On Physical Interactions & the World of God". 2008; www.ub.bw/news.cfm?t=905 8-5-12

Okulov Iu.I. Problem of Magnetic Charge & Neutrino Mass. 17th International Cosmic Ray Conf. Paris, July 13-25, 1981.

Penfield, Wilder (1975).,The Mystery of the Mind (Princeton: Princeton University Press, p. xiii. Wilder Penfield, *The Physical Basis of Mind*, ed. P. Laslett (Oxford: Basil Blackwell, 1950), p. 64.

Penrose, R. (1989). The Emperor's New Mind. Oxford University Press, Oxford.

Penrose, R. (1994). Shadows of the Mind. Oxford University Press, Oxford.

Pessa, E., and Vitiello, G. (2003). Quantum noise, entanglement and chaos in the quantum field theory of mind/brain states. Mind and Matter 1, 59-79.

Popp FA, Nagl W, Li KH, Scholz W, Weingärtner O, Wolf R. Biophoton emission. New evidence for coherence and DNA as source. Cell Biophys. 1984 Mar;6(1):33-52.

Popper, K. R. (1959). *The Logic of Scientific Discovery*. London: Hutchinson.

Popper, K. R., and Eccles, J. C. (1977). *The Self and Its Brain*. (Springer, Berlin, Heidelberg).

Hameroff, S.R., and Penrose, R. (1996). Conscious events as orchestrated spacetime selections. Journal of Consciousness Studies 3(1), 36-53.

Jibu, M., and Yasue, K. (1995). *Quantum Brain Dynamics and Consciousness*. Benjamins, Amsterdam.

Rescher, N. 2000, Process Philosophy – A Survey of Basic Issues, pp. 30-32, University of Pittsburgh Press, Pitssburgh, PA.

Ricciardi, L.M., and Umezawa, H. (1967). Brain and physics of many-body problems. Kybernetik 4, 44-48.

Salin, M. L., & Bridges, S. M. (1981). Chemiluminescence in wounded root tissue: Evidence for peroxidase involvement. Plant Physiology, 67, 43–46.

Shimizu, E., Y.-P. Tang, C. Rampon, J. Z. Tsien, *Science* **290**, 1170 (2000).

Skrbina, D. (2003). Panpsychism in Western philosophy. Journal of Consciousness Studies 10(3), 4-46.

Slawinska, D., Polewski, K., & Slawinski, J. (1992). The stress-induced electromagnetic emission from biosystems: Chemilumiescence response of plants to mechanical and chemical damage. Bioelectrochemistry and Bioenergetics, 28, 482–488

Smart, J.J.C. (1963). Philosophy and Scientific Realism. Routledge & Kegan Paul, London.

Stapp, H.P. (1993). A quantum theory of the mind-brain interface. In Mind, Matter, and Quantum Mechanics, Springer, Berlin, pp. 145-172.

Stapp, H.P. (1999). Attention, intention, and will in quantum physics. Journal of Consciousness Studies 6(8/9), 143-164.

Star, E. N., Kwiatkowski, D. J. & Murthy, V. N. (2002) Rapid turnover of actin in dendritic spines and its regulation by activity. *Nature Neuroscience*. *5*, 239-246.

Stephan, A. (1999). Emergenz. Dresden University Press, Dresden.

Stuart, C.I.J., Takahashi, Y., and Umezawa, H. (1978). On the stability and non-local properties of memory. Journal of Theoretical Biology 71, 605-618.

Stuart, C.I.J., Takahashi, Y., and Umezawa, H. (1979). Mixed system brain dynamics: neural memory as a macroscopic ordered state. Foundations of Physics 9, 301-327.

Taylor, J. Not a Chimp: The Hunt to Find the Genes that Make Us Human. Oxford, UK: Oxford University Press, 2009.

Kaszniak, A.W. and Scott, A.C., Cambridge, MA: MIT Press, pp. 507-540 (1996)

Van Wijk, R., Tilbury, R. N., Slawinski, J., Kochel, B., Gu, Q., Popp, F. A., Lilius, E.-M., Marnila, P., & van Aken, J. M. (1992). Biophoton emission, stress and disease: A multi-author review. Experientia, 48, 1029–1102.

Vitiello, G. (1995). Dissipation and memory capacity in the quantum brain model. International Journal of Modern Physics B 9, 973-989.

Vitiello, G. (2001). My Double Unveiled. Benjamins, Amsterdam.

Vitiello, G. (2002). No Matter, Never Mind: Dissipative quantum brain dynamics, eds. K. Yasue, M. Jibu and T. Della Senta, Benjamins, Amsterdam, pp. 43-61.

Wang, I. Bókkon, J. Dai, I. Antal, Spontaneous and visible light-induced ultraweak photon emission from rat eyes, Brain Res. 1369 (2011) 1Ḣ9.

Whitehead, A.N. (1978). Process and Reality. Free Press, New York.

Wilber, K., (2000) p 17; A Brief History of Everything, Shambhala, Boston, MA.

Winkler, K. et al. (2006), Nature 441, 853-856 (2006).

Addendum

PERIODIC TABLE OF PERSONALITIES

Introduction

Personality (from the Latin *persona* meaning mask) is a byproduct of **personhood**, a dynamic set of personal characteristics influencing motivations, cognitions and behaviors in different contexts. In the ancient Latin theater, the mask did not disguise the identity of a character; rather it represented or typified that character. As part of a community, the person is a socialized human being nurtured into and by the community norms and mores. A systematization of personality types is attempted based on four existential limits of personhood. Each limit has its own set of three different functions reflecting a unity of a three personality dimensions - spiritual, mental and physical. A correspondence with the views on life in ancient Scriptures of both the East and West seem to emerge, implying social stratification via personality types. Various functions combine to give 81 different types which can be arranged systematically into Groups & Periods. Group 1A is the Base Group and Group 8 is the Noble (Mahatma) Group. A multi-item inventory of traits can be developed to assess individual personality type based on the following premises:

1. Personhood is a tri-unity of: inner-self (pneuma), mind (psyche) and Body
2. There are four existential limits of personality, two of State, Self/ Super-self; and two of Status, Sovereign/Servant.
3. Each existential limit has a different set of three functions **co**rresponding.
4. Thus 81 different personality types exist, arranged systematically into Groups & Periods**.** Group 1A is the Base Group and Group 8 is the Noble Group.
5. A multi-item inventory of traits can be developed to assess individual personality type.

The "**self**," is both captivating and challenging topic in the social and biological sciences. Physical sciences are also deeply involved, ever since the Copenhagen interpretation of quantum mechanics (QM), rather *Copenpagan incantation),* because a mystical consciousness was brought into QM from occult connections. The body is not the self, but the self needs to be identical to the body. Otherwise how can it be **self**? While both are equally known, the self is a hidden physical reality which invisibly binds to and permeates the visible physical body. The former then is made from non-electric matter or *dark-matter* via its chemistry, in the same manner as the latter is made of electric matter or *light-matter* and its chemistry. The resonance between the two may be the basis for self-awareness. The **inner-man = the neutrino body (psyche) + the axion body (pneuma).** The **outer-man = the body. That is** verisimilar to the **Id, ego, superego** constellation. The "**self**" thus is identical to the "**body**" that **possesses it and is permeated by and integrally coupled to it.**

Contrary to **ordinary materialism**, this inner world of ***self***, or *interiority*, does not "*arise*" as an epiphenomenon out of the **ordinary matter** in the 'the learning, remembering **brain** encoded with representational maps.' The ***reality of interiority*** of the **self** cannot be different from the ***reality of the body*** itself. Else, **self** is a misnomer. **Self** cannot be any more *potential* than the **body**. The two are identical twin **realities** *co-created* at the very moment of conception. Both exist in the physical realm of matter, with this difference: one is of *light-matter* and the other of *dark-matter.*

Person/personality are delineated pretty differently both interculturally and intraculturally. They are often associated with metaphysical and mystic concepts based on *a priori* assumptions. **Self** presents itself in the *physical* realm as a dynamic monological and dialogical construct and needs be explored in the physical context. The I-me monologue is just as real as the I-you dialogue. "**Self**" exists not as a reification of an inert attribute, but as a dynamic unity of two ***totalities*** one belonging to the visible, temporal, external, physical, electromagnetic body and the other to its invisible nonetheless identical, enduring and internal, ***physical*** twin which is the **inner man**. The invisibility or undetectability of the latter points to non-electric matter.

Thus **personhood is** a physical composite of **body and self.** It is logically tied to the concept of human being and cannot be mere mystic intuitions or subjective experiences about what it is that makes a person. The invisible nonelectric, non-entropic ***self*** retains the durable genetic nature, while the visible entropic body/brain is subject to nurture. The former has the root of creativity of the genius as well as the destructivity of the criminal. The personhood and the personality together constitutes the total psyche.

Multiple theoretical models have been posited for the structure of personality. They offer diverse views on the expressions of personality and their assessments. These efforts reflect culturally loaded philosophic orientations as divergent as Western materialism and Eastern idealism. Humanistic and evolutionary schemes dominate most of the Western studies, while self-actualization towards the divine life is mostly the hub and nub of Indian Psychology. In the Western trait psychology, personality types are described mainly by using three factors, five factors, and sixteen factors. Spirituality is an elusive concept for Western behavioral science. Though Zohar and Marshall (2002) first used the term Spiritual Quotient (SQ), it cannot be measured. On the other hand, Indian psychology describes personality in 1 terms of spiritual qualities as 'guna' theory of sattva (divinity), rajas (dynamics), and tamas (delusion). Tamas is at the lower end of guna theory. In Ayurveda, personality is described in terms of combination of *tridoshas* or the three humors or controls: vata (wind), pitta (fluid), and kapha (lubrication), the most fundamental and characteristic principle of Ayurveda. These terms are difficult to translate, they are all-inclusive and multi-meaning.

The Indian concept of personality is a synthesis of the individual's mundane self and higher levels of mind. Matter, life and mind are all component units of infinite reality. The self though confined by finite realities can achieve progressive enlargement into the infinite, as an outcome of the incessant process of self-actualization and realization of goals aided by divine (yoga, meditation) and non-divine (karma) energy forces within the individual life. This involves a struggle of opposites for the survival of the self: birth vs death, or mortality vs. immortality, reality vs. unreality, good vs. evil, love vs. hate, in the Maya-world of experience. Perfection at the human level is an achievable assignment to be accomplished by conscious efforts. The genuinely human person pursues perfection with religious devotion and sincerity. 'Self ' is a part of the universal Self (Brahman); thus all are parts of the Supreme Reality. Everyone must realize this universal truth. Sri Aurobindo (*Letters on Yoga*, p. 220) describes a pluridimensional view of the person as a unity of two sets of three concentric rings: outer being, inner being and inmost being each of which has a mental, vital and physical component. A person is also portrayed as a triune vertical hierarchy of the inconscience, subconscious and superconscience. All dissonance between these parts is overcome by appropriate Yoga. Personality typologies are customarily based on body (sharira) and quality (guna) as integral units of personhood.

The Indian views are also invariably intertwined with the four stages or Ashrams of human life: Student (Brahmacharya), Householder (Grahastha), Forest Retiree (Vanaprastha), Forest Ascetic (Sannyasin). These are also associated with four stages of learning: unconscious incompetence, conscious incompetence, conscious competence, and unconscious competence. **Student** *serves* his teacher (guru means darkness remover) and darkness (*gu*) of unconscious incompetence is removed (*ru*) by devoted study of the Vedas. Student/learning has hereditary components, **The Householder *rules*** the house and leads a community (beginning with marriage) according to the *Laws of Manu* with ***kingly*** grace. This stage **The Forest Retiree (**beginning after the arrival of the first grandchild) tackles the question, what is ***man*** and comes to terms with one's humanity**. The Forest Dweller or Ascetic** begins by leaving home to carry out a ***spiritual*** existence.

Thus, in Indian Psychology personality dimensions are related to servant, sovereign, human and spiritual aspects of life. Indian authors Daftuar et al (2002) reports 15 dimensions related to spirituality, with 99 items. The dimensions include God, religiosity, Soul, Consciousness, Spirituality, Values, Fate, Karma and Divinity. They are rooted in the ancient Hindu Scriptures.

Current Western perspective of personality is secular and materialistic with an evolutionary motif, though the Bible presents a unity of three dimensions of personhood-spirit (wind), soul (mind) and body (flesh)- each with four boundaries (divine, human, servant and sovereign), which has overtones similar to the Vedic view described above. The Hebrew Scriptures present the Branch (Messiah) as the Servant-King-God-Man. The four Gospels respectively present the Messiah as King, Servant, Human and Divine. Matthew takes the genealogy of the Messiah to King David, Mark omits it altogether for a servant, Luke takes it all the way to the man Adam and John omits it altogether for the Deity. Human and divine aspects are intrinsic traits; servant and king aspects are extrinsic traits. The Biblical views do not find a place in modern Western psychology. This is in sharp contrast to the heart of Eastern philosophy which regards the dimensions of spirit (consciousness), mind (intellect), and body relations as integral parts of the whole person.

Personality results from personhood. Personhood is a tri-unity of sentience (life), mind (self) and sapience (inner-self), which is traditionally depicted as body/sensorium, mind/brain and heart/spirit. The "self is not the same as the "body". It cannot be categorically different from the body either. "Self" and the "body" are identical *physical* twins co-created at the moment of conception, except that one is visible made of "electric" matter (*light-matter*) with mass and the other made of non-electric matter (*dark-matter*) with negligible mass. Person/personality are delineated pretty differently both interculturally and intraculturally. They are often associated with metaphysical and mystic concepts based on *a priori* assumptions. **Self** presents itself in the *physical* realm as a dynamic monological and dialogical construct and needs be explored contextually i.e. *physically*. The I-me monologue is just as real as the I-you dialogue. **"Self"** exists not as a reification of an inert attribute, but as a dynamic unity of two ***totalities*** one belonging to the visible, temporal, external, physical, electromagnetic body and the other to its invisible nonetheless identical, enduring and internal, ***physical*** twin which is the **inner man**. A **neutrino body** and an **axion body** both permeating and coupled to each other and to the visible body will have enormous explanatory powers. The **inner-man = the neutrino body (psyche) + the axion body (pneuma).** The **outer-man = the body. That is** verisimilar to the **Id, ego, superego** constellation. The "**self**" thus is identical to the "**body**" that **possesses it and is permeated by and integrally coupled to it.**

An objective methodical approach to the study of personality and its types is possible, if existential limits of ‘operation’ can be defined. Only four such operational limits of personhood exist. These are: Super-self (Mahatma), Human Self (Natural), Sovereign, Servant. The first two relate to the intrinsic human State or condition and the last two relate to the human Status (attainment). States are fixed and inherited. Statuses are acquired. The Vedas view human beings as composed of both matter and spirit. Our physical body, mind/intellect are the matter components and the soul is the spirit - the real Self and the driving force of the physical components. An invisible dark matter body coupled to the visible human body has been posited by Benjamin (2003, 2007A&B).

Then each of the four existential extremities is defined by its own set of three qualities which are functionally related to body, mind and spirit. In the Biblical sense it is a tri-unity of spirit, mind and body (in that order). Empirically, a vegetative state is universal for all sentience. An animal state is common to animals and humans. Human state stands unique as what Prof. Hauser (2008) calls ‘humaniqueness’. The body acts, mind thinks and the spirit motivates. The three aspects of each State and Status are as follows:

1. Super-self State: Spiritual (S), Mental (M), Natural (N).
2. Self State: Thinking (T), Feeling (F), Acting (A) [TFA of Hutchins, 1979].

3. Sovereign Status: Authoritative, Authoritarian, Permissive
4. Servant Status: Compliant, Assertive, Aggressive

Combination of all these factors result in 81 personality types as shown below. They may be arranged in the form of a Periodic Table of Personalities as Groups and Periods. This should cover the full spectrum of existential possibilities of personality types, both normal and abnormal. An assessment scale for normal personality types is suggested. Such assessments help to summarize, predict, and explain a person's attitudes and behavior over all.

Personality Types

Thurstone (1934) and Eysenck (1991) discuss the vectors and dimensions of mind. Myers - Briggs' 16-type Personality Indicator is based on Jung's four categories of people with two types in each (Myers & McCauley, 1985). David Kersey (1978, 1995, 1998) randomly arrives at 16 different personality configurations which he claims correspond to those of Myers. Kersey considers brain to a computer with temperament as its physical base or hardware with character as its software. Then, temperament is a configuration of pre-dispositional inclinations and character is an array of dispositional habits emerging as individual attitudes and actions from the physical base. Personality is a complex configuration of functions pertaining to both nature and nurture (Coan 1994; Schwartz, et al 1995; Oldham & Morris 1991). Setting natural limits to these functions, may significantly help to systematize the relationships of personal inclinations and predispositions. *Self * is a limiting singularity of personhood. At its zenith is the Super-self which is another ultimate state. It is not merely an attribute or claim or the Freudian Super-Ego, but the god-man, 'mahatma' as it were. To be sovereign is another limit and at its nadir is the lowliest servant. These are finalities that depict finite human bounds or limiting tendencies.

Thus, there are four "archetypical' bounds within which people exist and function; two are *state* and the other two *status* limits. Super-self and Self are *state* limits. Sovereign and Servant are *status* limits. State and Status are both natural endowments, though status is more amenable to nurture. Each of these existential limits, has three possible types of dispositions. The Super-self manifests in one of three coequal levels: 1.Inner-self/Spiritual (S), 2. Mental (M)/Soulish, 3. Outer-Self/Natural (N). The Self operates predominantly in one of three modes: 1. Thinking (T) 2. Feeling (F). 3. Acting/Acting (A). Hutchins (1979) has discussed TFA. The Sovereign status exercises control in one of three ways: 1. Authoritative (U) 2. Authoritarian (R) 3. Permissive (P). The Servant status prefers one of three styles: 1. Compliant (C). 2. Assertive(V) 3. Aggressive (G). All these are familiar terms in psychology. . Each individual personality is a composite configuration of these limiting predispositions and propensities. Table 1 summarizes the four categories and twelve types.

Table 1

STATE/STATUS		TYPES	
1. Super-self	i. Spiritual (S)	ii. Mental (M)	iii. Natural (N)
2. Self	i. Thinking (T);	ii. Feeling (F)	iii. Acting(A)
3. Sovereign	i. Authoritative (U)	ii. Authoritarian (R)	iii. Permissive (P)
4. Servant	i. Compliant (C);	ii. Assertive (V);	iii Aggressive (G).

These constitute eighty-one different combinations of personality types as in Table 2: The types in the Super-self and Self are paired giving, ST, SF, SA; MT, MF, MA; NT, NF, NA. Similarly, types in the Sovereign and Servant combine to yield, PG, PV, PC; RG, RV, RC; UG, UV, UC. Each of the 9-pairs in one set combines with each in the other set to give 81 types of personalities that should cover the whole spectrum of normal and abnormal types. Thus, NA will have the following 9-types. NAPG, NAPV, NAPC, NARG, NARV, NARC, NAUG, NAUV, NAUC. This is the Base group. The 9-types under ST is the Noble group: STPC, STPG, STPV, STRG, STRV, STRC, STUG, STUV, STUC. All the other types fall in between. with a total of 81 types. These form 8 major groups with (Groups 1A & 1B) and 9 periods. The numbers in brackets below each Type refer to the product of values assigned to the functions as in Table

5 . According to Oldham (1994), the personality disorders are extreme manifestations of normal personality traits and behaviors. Millon & Davis (1996) holds a similar view.

Table 2 PERIODIC TABLE OF PERSONALITY TYPES

ST	SF	SA	MT	MF	MA	NT	NF	NA	PG	PV	PC	RG	RV	RC	UG	UV	UC
[9]	[6]	[3]	[6]	[4]	[2]	[3]	[2]	[1]	[1]	[2]	[3]	[2]	[4]	[6]	[3]	[6]	[9]

	IA	IB	II	III	IV	V	VI	VII	VIII
1.	NAPG	NFPG	NTPG	MAPG	MFPG	MTPG	SAPG	SFPG	STPG
	[1]	[2]	[3]	[2]	[4]	[6]	[3]	[6]	[9]
2.	NAPV	NFPV	NTPV	MAPV	MFPV	MTPV	SAPV	SFPV	STPV
	[2]	[4]	[6]	[4]	[8]	[12]	[6]	[12]	[18]
3.	NAPC	NFPC	NTPC	MAPC	MFPC	MTPC	SAPC	SFPC	STPC
	[3]	[6]	[9]	[6]	[12]	[18]	[9]	[18]	[27]
4.	NARG	NFRG	NTRG	MARG	MFRG	MTRG	SARG	SFRG	STRG
	[2]	[4]	[6]	[4]	[8]	[12]	[6]	[12]	[18]
5.	NARV	NFRV	NTRV	MARV	MFRV	MTRV	SARV	SFRV	STRV
	[4]	[8]	[12]	[8]	[16]	[24]	[12]	[24]	[36]
6.	NARC	NFRC	NTRC	MARC	MFRC	MTRC	SARC	SFRC	STRC
	[6]	[12]	[18]	[12]	[24]	[36]	[18]	[36]	[54]
7.	NAUG	NFUG	NTUG	MAUG	MFUG	MTUG	SAUG	SFUG	STUG
	[3]	[6]	[9]	[6]	[12]	[18]	[9]	[18]	[27]
8.	NAUV	NFUV	NTUV	MAUV	MFUV	MTUV	SAUV	SFUV	STUV
	[6]	[12]	[18]	[12]	[24]	[36]	[18]	[36]	[54]
9.	NAUC	NFUC	NTUC	MAUC	MFUC	MTUC	SAUC	SFUC	STUC
	[9]	[18]	[27]	[18]	[36]	[54]	[27]	[54]	[81]

The 9 rows are the Periods PG, PV, PC, RG, RV RC, UG, UV, UC.

The 9 columns are the Groups NA, NF, NT, MA, MF, MT, SA, SF, ST.

The numbers in brackets represent quantification of personality types

Periods: Status Functions See Table 3

The horizontal groups reflect the changing lifespan development of personality. Eric Erickson's (1963, 1968, 1980, 1986) eight stages seem most appropriate here. Period 1 is PG Permissive/Aggressive' from birth to one year when the child may act with Trust Vs. Mistrust. Period 2 covers ages 1-3 when the child is ***PV* with** a 'Permissive/Assertive' Autonomy vs. Shame/Doubt. Period 3 ***PC* is** 4-5 years 'Permissive/Compliant' "Initiative Vs. Guilt". Period 4 ***RG*** with 'Authoriatarian/Aggressive' 6 years to puberty with 'Industry vs. Inferiority. Period 5a and 5b is RV/RC is 'Authoritarian/Authoritative/ Compliant' Adolescence with Identity Vs. Role Confusion. Period 6 is ***UG*** 'Authoritarian/Aggressive young adulthood of Intimacy Vs. Isolation. Period 7 is ***UV*** 'Authoritative/Assertive' of middle adulthood, the settled life of Generativity vs. Stagnation. Period 8 is ***UC*** 'Authoritarian/Compliant' late adulthood of Integrity Vs. Despair.

Groups: State Functions. See Table 4

Group I A is the Base group and Group VIII is the Noble group

The vertical groups reflect the more permanent states of inherited traits, including evaluative attitudes of conscience which have been deliberately ignored by modern psychology. Jane Loevinger (1994) asked: " Has Psychology Lost Its Conscience? Novelist Kurt Vonnegut as quoted by Alfred Kantor (1971), proposed a basic unit of "conscience measurement, to be called the "Stowe" in honor of the "only writer in history who had an effect on the course of world affairs. Friedenberg, E. Z.,& Havighurst, R. J.(1948) attempted to measure strength of conscience. Pittel and Mendelsohn (1969), developed situational appraisal inventory and validation of a measure of evaluative attitudes. Farr A. Curlin et al (2007) discussed the need to incorporate conscience into personality assessments. Daftuar et al (2002) have introduced conscience as a function in personality types. **S**everal States in the U.S.A have passed right of conscience acts to protect health care providers from adverse consequences for non-participation in medical services that would violate their consciences (See Vischer 2006). Julian Savulescu (2006) raised the issues of conscientious objection in medicine. Conscience, childhood innocence (adult gullibility) and life-span deceit, howsoever ill defined, are built-in parameters of the conscious psyche shaping the human personality.

The groups correspond to the developmental stages of Piaget (1971) and Kohlberg (1969}. Group IA is Sensorial Stage of Innocence, birth- 2 years, Piaget's Sensory-Motor Stage. Group IB is Predispositionary Stage of Deceit (2-7 yrs), Piaget's Pre-operational Stage. Group II is Guilt-validation Stage of Conscience, (7-11 yrs), Piaget's Concrete Operation. Group III- VIII refers to 11 years and up, Piaget's Formal Operations and Kohlberg's (1969) Pre-conventional Morality. Group III is Kohlberg's Stage 1: Obedience and Punishment Orientation. Group IV is the Stage 2 Individualism and Exchange: sorting out different views of different individuals on the same issues, including whatever view is handed down by authority figures. Group V is the Stage 3 of Kohlberg's Conventional Morality: the tension of living up to the standards of the family and community and establish good interpersonal relationships. Group VI is Kohlberg's Stage 4. Maintaining the Social Order. Rationale for intimacy in bilateral relationships with family members or close friends, and be helpful in the context of the laws and relationships of society at large. Group VII is Kohlberg's stage 5 of Social Contract and Individual Rights. Group VIII is Kohlberg's Stage 6 of Universal Principles for a better world order by protecting individual rights and settling disputes peacefully and freely.

Natural (N), Mental (M) and Spiritual (S) functions with each coupled to Acting (A), Feeling (F), Thinking (T) functions form nine groups. These coupled functions exist across the life span periods. Progression from the nadir Base group NA to the zenith Noble group ST depends on the efforts of each person. These groups may be subdivided as follows: 1. The Noble or Spiritual set consisting of the groups ST (VIII), SF (VII) and SA (VI). 2. The Alert Mental set made of the groups MT (V), MF (IV) and MN (III)

3. The sensory Natural set comprising the groups, NF (1B), and NT (II). 4. The stimulus-response primal Base group NA (1A). This is the basic group. Progression from one state to the next state is possible by efforts. In the traditional Indian Psychology these groups are identified as permanent life stations or castes of Brahman (priest, intellectual), Kshatriyas (warrior, ruler), Vaishya (artisans, traders) and Shundra (worker, servant) which can change after death only in a deterministic manner by the karma (efforts) of each station. These are also called varnas (varna = color). They are primal types originating from different parts of the Primordial Man (*Purusha*). Each varna has hundreds of predetermined gradations which correspond to individual aptitudes, attitudes and vocational preferences in life. These castes reflect the social stratification or personality types. This is supported also in the *Republic* of Plato, where the ideal society consists of Philosophers who rule through wisdom, Warriors, and the masses. Ancient Rome and Greece had such stratifications of personality types. One's caste corresponds to one's natural aptitudes and orientations for a vocation in life, not necessarily a fatalistic socially-fixed operational schema.

Table 3 Periods Status Functions of Personality

Periods & Characteristics	Erickson Stages
1. **PG:** Permissive-Aggressive	Trust vs. Mistrust
2. **PV:** Permissive -Assertive	Autonomy vs.Shame
3. **PC:** Permissive-Compliant	Initiative vs.guilt
4. **RG:** Authoritarian-Aggressive	Identity/Role Confusion
5. **RV:** Authoritarian-Assertive	Industry vs. Inferiority (Contd.)
6. **RC:** Authoritarian-Compliant	Industry vs Inferiority
7. **UG:** Authoritative-Aggressive	Intimacy vs. Isolation
8. **UV:** Authoritative-Assertive	Generativity/Stagnation
9. **UC:** Authoritative-Compliant	Integrity vs. Despair

Table 4 State Functions of Personality

Groups	Piaget/Kohlberg Stages
I **NA:** Natural/ Acting	Sensorial, Dependent Impulsiveness
IB **NF:** Natural/Feeling.	Sensorial, Sensitive, Preoperational Innocence
II **NT:** Natural /Thinking.	Concrete Operational
III **MA:** Mental/Acting.	Formal Operational/Obedience Penalty
IV **MF:** Mental/Feeling	Individualism and Exchange

V **MT:** Mental/Thinking Conventional Morality
VI **SA:** Spiritual/Acting Maintaining the Social Order
VII **SF:** Spiritual/Feeling: Social Contract and Individual Rights
VIII **ST:** Spiritual/Thinking Universal Principles

Quantification of Personality Styles

A numerical value is obtained for each configuration in the table by assigning values arbitrarily 1, 2 & 3 to each function in the four categories, as in Table 5.

Table 5

Super-self	Self	Sovereign	Servant
Spiritual(S) = 3	Thinking(T) = 3	Authoritative(U) = 3	Compliant(C) = 3
Mental(M) = 2	Feeling(F) = 2	Authoritarian (R) = 2	Assertive (V) = 2
Natural(N) = 1	Acting(A) = 1	Permissive(P) = 1	Aggressive(G) = 1

Equal values in **Table 2 & 5** do not mean identical types. They reflect the balancing influence of the state functions on the status functions.

Individual Inclination Check List:

Table 6 has 2 sets of 3 **state functions** each, and each function has nine items. Similarly **Table 7** similarly describes two sets of **status functions**. Check the quality that seems most befitting you. Answer as fast as possible across the table. First impression is the best indicator. Count the checks in each column and circle the alphabet at the bottom of the column with the highest score. Write the four alphabets below. This is your personality code.

For example suppose an adult gets [N] [A] [P] [G] with a numerical value 1. NA is infantile/sensory, PG is permissive king/aggressive servant. NAPG is an immature adult given to impulsive gratifications, suggestive of pathology, a *Base Mind/Durjan*. On the other hand, if a 65 year old gets [S] [U] [T] [C] with a numerical value of 81. SU is (Super-self)/Authoritative, TC is Thinking/compliant servant. That is a *Noble Mind/Mahatma*.*

* Note: In the East generally Noble Mind/Mahatma is thought to be attained through self-realization and deeds without attachments. In the West generally divinely infused or induced Regeneration/Rebirth leads to continuous renewal of mind.

Table 6 State Items Two sets of Three Functions

	Set 1			Set 2	
Spiritual (S)	Mental (M)	Natural (N)	Thinking (T)	Feeling (F)	Acting (A)
Altruistic	Artistic	Athletic	Alert	Emotional	Compulsive
Authentic	Analytical	Cunning	Attentive	Engaging	Vigorous
Caring	Autonomous	Critical	Evaluative	Ethical	Flamboyant
Conscientious	Curious	Egocentric	Inventive	Dependent	Impulsive
Contemplative	Disciplined	Instinctive	Inquiring	Idealistic	lively
Empathetic	Factual	preoccupied	original	Indulgent	natural
Honest	Intuitive	Programmed	Perceptive	Sensitive	Practical
Insightful	Objective	Reciprocal	Skeptical	Sympathetic	Reflexive
Wise	Self-respect	Subdued	Systematic	Volatile	Reactive
S	M	N	T	F	A

Table 7. Status Items Two Sets of Three Functions

	Set 3			Set 4	
Authoritative	**Authoritarian**	**Permissive**	**Assertive**	**Aggressive**	**Compliant**
U	R	P	V	G	C
Calm	Conceited	Carefree	Confident	Demanding	Easygoing
Creative	Officious	Unsure	Expressive	Doctrinaire	Informal
Deliberate	Formal	Laid-back	Judicious	Gritty	Ministering
Didactic	Inflexible	Lenient	Manageria	Private	Nurturing
Helpful	Insecure	Liberal	Positive	Reactive	Unpretentious
Poised	Probing	Indifferent	Pragmatic	Reserved	Relaxed
Realistic	Pessimistic	Optimistic	Proactive	Risky	Responsive
Serene	Smug	Retiring	Straight	Scheming	Undemanding
Virtuous	Cynical	Tolerant	Tactful	Strongwille	Witty
U	R	P	V	G	C

Addendum References

Benjamin, Philip.

Bio Dark-Matter Chemistry", International Journal of Current Research and Reviews Vol 4 issue 20, 2012

Dark Chemistry & the Paranormal, WorldComp'10 Proceedings of the International Conference on Artificial Intelligence, Vol II, pp 633-39, July 12-15, 2010, CSREA Press, Las Vegas, NV.

Dark chemistry & paranormal phenomena, International Journal of Applied Science & Computations, Vol. 17, No. 1 Pages 16 to 36, June 2010

Mind Matter, Noetic Journal Vol 4 # 4, 351-360, 2003 [Nobelist Sir John Eccles Centennial Edition].

Dark Chemistry or Psychic Spin Pixel? NeuroQuantology, June 2007, Vol. 5 # 2, 197-204.

Dark Matter & Dark Chemistry NeuroQuantology September 2007, Vol 5 # 3, 322-326. http://noeticcenter.tripod.com noeticcenter.tripod.com/blog/

KARL JASPERS FORUM TA91 (Müller) Commentary 5 (to R2) COMPLETE EXPERIENCE/ Dark Matter by Philip Benjamin 23 December 2006, posted 30 December 2006 http://www.kjf.ca/91-C5BEN.htm

Coan, R. W., (1994). Personality types. In Corsini, R. J., (Ed.), *Encyclopedia of Psychology.* 2nd ed., Vol. 3 (pp. 58-60). New York: Wiley.

Curlin, Farr A., Ryan E. Lawrence, Marshall H. Chin,, and John D. Lantos, Religion, Conscience, and Controversial Clinical Practices, New England Journal of Medicine, February 8, 2007, Volume 356:593-600, Number 6.

Daftuar, C. N., C. Dasgupta, C., & Nira Manghrani (2002), National Conference on Yoga and Indian Approaches to Psychology Pondicherry, India, September 29- Oct.1, 02

Erickson, E.H., (1963). Childhood and Society (2nd ed.). New York, Norton [9,10] Erickson, E.H., (1968). Identity: Youth & Crisis, New York, Norton [10]

Erickson, E.H., (1980). Identity & the Life Cycle, New York, Norton [10]; Erickson, Erickson, J.M., and Kivnick, H.Q. (1986). Vital Involvement in Old Age: The Experience of Old Age in Our Time, New York, Norton [10].

Eysenck, HJ. (1991). Dimensions of personality: 16: 5 or 3? criteria for a taxonomic paradigm. Personality and Individual Differences, 12, 773-90.

Friedenberg, E. Z.,& Havighurst, R. J.(1948). An attempt to measure strength of conscience. Journal of Personality,17,232-243 content.apa.org/journals/ccp/33/4/396

Hauser, Marc. (2008), A theory of "humaniqueness‘, The annual meeting of the American Association for the Advancement of Science, February, 2008

Hutchins, David E. (1979) Systematic Counseling: The T-F-A Model for Counselor Intervention. (EJ202833) Personnel and Guidance Journal, v57 n10 p529-31 Jun 1979

Kantor, Alfred. (1971) The Book of **Alfred Kantor**, McGraw-Hill, NY.

Keirsey, David. (1995). *Portraits of Temperament.* 3rd. ed. Del Mar, CA: Prometheus Nemesis. pp. 6,12;

Keirsey, David and Bates, Marilyn. (1978). *Please Understand Me: Character and Temperament Types.* Del Mar, CA: Prometheus Nemesis, pp. 3-4, 29-30.

Keirsey, David. (1998) *Please Understand Me II.* CA: Prometheus Nemesis, p. 26. Jane

Kohlberg, L. (1969). Stage and sequence: The cognitive developmental approach to socialization. In D. Goslin (Ed.), *Handbook of socialization theory and research.* Chicago: Rand McNally.

Loevinger, Jane (1994) Journal of Personality Assessment, Volume 62, Issue 1 February 1994, pages 2 – 8, DOI: 10.1207/s15327752jpa6201_1

Millon, T. & Davis, R., (1996). *Disorders of Personality: DSM-IV and Beyond.* 2nd ed. New York: Wiley.

Myers, I. B., & MCCauley, M.H. (1985). A guide to the development and use of the Myers-Briggs Type Indicator. Palo Alto, CA: Consulting Psychologists Press.

Oldham, J. M. & Morris L. B., (1991). *The Personality Self-Portrait: Why You Think, Work, Love, and Act the Way You Do*. New York: Bantam.

Oldham, J. M., DSM IV, 1994 pp. 417-432.

Piaget, J. (1971). The theory of stages in cognitive development. In Proceedings of the CTB/McGraw-Hill conference on ordinal scales of cognitive development (D. R. Green, M. P. Ford, & G. B. Flamer, (Eds), *Measurement and Piaget Proceedings* (pp. 1-11). New York : McGraw-Hill. BF723 .C5 C2 1969

Pittel, Stephen M.; Mendelsohn, Gerald A (1969), Situational Appraisal Inventory: Development and validation of a measure of evaluative attitudes, Journal of Consulting and Clinical Psychology. 1969 Aug Vol 33(4) 396-405

Savulescu J. Conscientious objection in medicine. BMJ 2006;332:294-297. BMJ 2006;332(7536):294 (4 February), doi:10.1136/bmj.332.7536.294

Schwartz, M. A., Wiggins, O. P. & Norko, M. A., (1995). Prototypes, ideal types, and personality disorders: the return to classical phenomenology. In Livesley, W. John, (Ed.), *The DSM-IV Personality Disorders, 1994.* New York: Guilford.

Thurstone, L. L. (1934). The vectors of the mind. *Psychological Review, 41*, 1-32.

Vischer RK (2006), Conscience in context: pharmacist rights and the eroding moral marketplace. Stanford Law Pol Rev 2006;17:83-119. [Medline].

Made in the USA
Las Vegas, NV
25 September 2026

55162846R00059